Charles Henry Stowell

A Healthy Body

A Text Book on Anatomy, Physiology, Hygiene, Alcohol, and Narcotics

Charles Henry Stowell

A Healthy Body
A Text Book on Anatomy, Physiology, Hygiene, Alcohol, and Narcotics

ISBN/EAN: 9783744670036

Printed in Europe, USA, Canada, Australia, Japan

Cover: Foto ©berggeist007 / pixelio.de

More available books at **www.hansebooks.com**

A HEALTHY BODY.

A TEXT-BOOK ON

ANATOMY, PHYSIOLOGY, HYGIENE, ALCOHOL, AND NARCOTICS.

FOR USE IN INTERMEDIATE GRADES IN PUBLIC AND PRIVATE SCHOOLS.

By CHARLES H. STOWELL, M.D.,

PROFESSOR OF HISTOLOGY AND MICROSCOPY, UNIVERSITY OF MICHIGAN; AUTHOR OF "STUDENT'S MANUAL OF HISTOLOGY;" "THE HUMAN TOOTH," "MICROSCOPICAL DIAGNOSIS;" "PHYSIOLOGY AND HYGIENE;" LATE EDITOR OF "THE MICROSCOPE," ETC.

𝔉ully 𝔈llustrated

WITH ORIGINAL SKETCHES BY THE AUTHOR.

THIRD EDITION REVISED.

[*Fiftieth Thousand.*]

SILVER, BURDETT & CO., PUBLISHERS

NEW YORK . . BOSTON . . . CHICAGO

1891

INDORSEMENT.

We have examined the school text-book, entitled "A Healthy Body," by Doctor CHARLES H. STOWELL.

As it faithfully teaches the nature of alcoholic drinks and other narcotics in connection with relative physiology and hygiene, and in language well adapted to grade, we heartily commend and indorse it for INTERMEDIATE or GRAMMAR GRADE pupils or schools.

MARY H. HUNT,

National and International Supt. Dept. of Scientific Instruction of the Woman's Christian Temperance Union.

Advisory Board for U. S. A.
ALBERT H. PLUMB, D.D.,
DANIEL DORCHESTER, D.D.,
Hon. WILLIAM E. SHELDON,
Rev. JOSEPH COOK.

University Press:

JOHN WILSON AND SON, CAMBRIDGE.

PREFACE.

THIS book has been written to meet the demand for the study of the human body in the intermediate and grammar grades of the public schools. Especial attention has been given to the effects of alcoholic drinks on the organs and tissues of the body. In this respect no statement has been made that is not capable of positive proof. The author has also strongly insisted on the injury to young men and boys caused by the habit of smoking tobacco, which in too many cases prepares the way for more disastrous habits. Good health is one of the safeguards not only against suffering, but also against crime. Vicious habits often result from a lack of knowledge of the laws of health, while both physical and mental health are frequently impaired by conditions which a proper understanding of hygiene would obviate. Therefore, special attention has been given to general hygiene. It is believed that the study of the human body, together with a knowledge of the effects of alcohol upon it, will do much to make the coming generation wise, temperate, and moral.

CHARLES H. STOWELL.

UNIVERSITY OF MICHIGAN,
December, 1888.

PREFACE TO THE THIRD EDITION

So short a time has elapsed since the first appearance of this book that radical changes are not required. Some slight alterations have been made in the text, a number of new illustrations substituted, and a list of questions added to each chapter. Grateful acknowledgment is made for kind and intelligent suggestions from Mrs. Mary H. Hunt, in the preparation and revision of the text on alcohol and other narcotics.

CHARLES H. STOWELL.

WASHINGTON, D. C.
August, 1891.

CONTENTS.

vi CONTENTS.

A HEALTHY BODY.

CHAPTER I.

WHAT IS ALCOHOL?

What is Alcohol? Alcohol is that part of all spirituous drinks which intoxicates. It is lighter than water, colorless, and has a sharp, stinging taste. It burns readily, gives very little light, and no smoke, but a great deal of heat. When taken into the stomach it acts as a powerful poison.

How it was discovered. It is not known when fermented liquors were first used. But about seven hundred years ago chemists (or alchemists, as they were then called) were continually trying to discover two things. One was to find something that would turn all the common metals, as iron, into gold, and thus make the discoverer very wealthy. Another object was to find something that would prevent sickness and death, bring back youth to the aged, and keep the young from ever growing old. When this last should be found, it was to be called "the elixir of life."

Was Alcohol this Elixir? It happened that about this time one of these chemists succeeded in distilling some alcohol from a fermented liquid. He tasted it, and at once noticed its power to excite the body and make him forget care and trouble. He felt young again, and was now sure he was the possessor of the long sought-for " elixir." He advised his friends to drink freely of the newly discovered liquid, telling them it would bring back to them their youth and strength. But, as might be expected, after a wild career of drunkenness he died in a drunken stupor.

How it is made. The solid parts of most ripe grains, as corn and barley, consist largely of starch. This starch can be changed into sugar, and then into alcohol and carbonic-acid gas, as described on page 14. In such ripe fruits as the grape and the apple we find sugar and water. If the juice of these fruits be pressed out and exposed to the sun and air, or be kept warm in any way, the sugar will be changed into two very different substances, — alcohol and carbonic-acid gas. While the juice is undergoing this change, bubbles may be seen coming to the surface ; this is the carbonic-acid gas rising and escaping. But the alcohol remains in the liquid.

What causes Sugar to change to Alcohol? The microscope has shown that there are minute bodies floating in the air, too small to be seen by the unaided eye, called " ferments." These ferments are very numerous, and of different kinds, and each kind does a different work. One kind of ferment enters any warm, sweetish liquid it may find, as the juice of fruits, and working on its sugar, or sweet principle, turns it to alcohol and carbonic-acid gas. Therefore we say that "ferments" cause the sugar to turn to alcohol.

A Universal Law. Fermentation entirely changes the character of the substance it works upon. This law that pervades all nature is seen in the case of fruits and grains. The process of fermentation entirely changes them. There is no juice in a ripe grape that will burn, neither is there anything in it that will in-toxicate. The grape is wholesome until it begins to decay. So also is its juice wholesome while it is in the ripe fruit; but it is quickly changed into a poisonous fluid after being pressed out. Milk is wholesome; but if it remain exposed to the sun or air until it becomes sour, it is then likely to produce disease if taken into the body. Beef is wholesome; but if we use it as a food after it has decayed, it will prove a serious experiment for us. So the juices of fruits are wholesome until they are pressed out and allowed to decay. Then fermentation sets in, and their character is entirely changed. Alcohol is not a natural part of either the unripe or the fully ripe fruit or grain.

The Alcohol Appetite. As in the case of opium and some other poisons, the use of alcohol in beer, wine, cider, or in any such liquor, excites an appetite for more alcohol. When we are thirsty, a small amount of water will satisfy us; and when we are thirsty again, no more water will be required to quench the thirst than before. But with alcohol it is different. One drink creates a desire for another; and we shall find in a later chapter that its effect on the tissues is such that this desire will steadily increase as the system comes more and more under its influence. If a small quantity satisfies at first, more will soon be required, until sooner or later nothing will quench the burning thirst until the whole body is made either stupid or insensible.

This appetite may lie quiet for years, only to be awakened by the accidental taste of some alcoholic drink. To escape from its power, all drinks that contain the smallest quantity of alcohol should be carefully avoided. Such light drinks as cider and home-made beer, and such dishes as brandy sauce and wine jelly, should be shunned. They contain a certain quantity of alcohol, and therefore cannot be taken without danger; for the general rule holds true, that one taste or drink of alcohol creates a desire for another.

How fast will it grow? There is a great difference in the degree of rapidity of the growth of this appetite. In many cases the appetite is so easily formed, and grows with such speed, that when once it fastens itself on its victim, it never releases its hold while life lasts. No one intends to become a hard drinker when he takes his first glass of beer or wine. As a rule, that is done thoughtlessly. It is the treacherous power that alcohol possesses of creating an appetite for its use, and the growth of this appetite, that lead so many to ruin.

Its Terrible Power. This power of alcohol to create an ever increasing appetite for its use is one of the most fearful things that can be said about it. To take the first glass of cider, beer, or wine is a dangerous venture ; the alcohol in it may excite an appetite which will eventually cost all one is worth in money, in body, and in mind for its gratification.

Whom will it affect? Although persons of a highly nervous temperament are the soonest and the most seriously affected, yet none are sure of escape. It affects persons of all ages, all temperaments, and in all conditions of life.

Heredity. The power of alcohol for evil is shown in the suffering it brings to the children of intemperate parents. As the faces and dispositions, the likes and dislikes, of the parents are present in their children, so the love for drink may descend from parent to child. When we have completed the study of the effects of alcohol on the organs and tissues of the body, and then recall the fact that the whole body may be brought under its influence, we shall not be surprised to know that its effects may be inherited as easily as any feature or characteristic of the body or mind. A young person may be entirely unaware that he has this desire for strong drink until, by the taste of some light drink, or some sauce flavored with wine, the sleeping appetite is aroused.

Alcohol causes such changes in the tissues of the

body that the children of intemperate parents are sometimes born without reason or judgment. In a report to
the Legislature of Massachusetts, Dr. Howe says that
three fourths of all the idiots in that State are the children of intemperate parents. The lives of the parents
and children are so closely linked that it may be stated
as a rule, to which there are but few exceptions, that
" where both parents are intemperate, their children will
have an appetite for strong drink ;" and " where one parent is intemperate, the children are likely to inherit a
love for strong drink." Children who inherit this love
for drink, who are born with the alcohol appetite, begin the work of life at a fearful disadvantage ; and it
will require great caution and courage on their part to
conquer the evil for which they are not responsible.

QUESTIONS.

1. What is alcohol? Name some of its qualities.
2. How was it discovered, and what followed?
3. What are found in such ripe fruits as grapes and apples?
4. How is alcohol made from the juice of ripe fruits?
5. What causes sugar to turn to alcohol?
6. What is the law about the results of fermentation?
7. How does fermentation affect various articles of food?
8. What changes the character of the juices that have been pressed
   out of fruits?
9. Is alcohol a natural part of fruits and grains?
10. How do water and alcohol differ in quenching the thirst?
11. How can we escape from the power of alcohol? Why?
12. What is one of the most fearful things about alcohol?
13. Whom will it affect?
14. Tell something about the hereditary effects of strong drink.

CHAPTER II.

FERMENTED LIQUORS.

THERE are two classes of liquors, known as the "fermented" and the "distilled."

What is necessary for Fermentation? In the last chapter we said that the changing of sugar into carbonic-acid gas and alcohol was one kind of fermentation. When corn or barley is used to make a fermented liquor, the dried grains are moistened with water and allowed to "sprout." When they begin to grow, their starch is changed into sugar. If we taste a kernel of corn after it has sprouted, we notice at once that it is sweet. Dry sugar will not ferment, neither will a mixture ferment that has in it a great deal of sugar. A moderate degree of heat is necessary for fermentation, for neither a very cold nor a very hot mixture will ferment. These facts show that at least three things are necessary to produce alcohol, — first, sugar; second, water; and third, heat. But a fourth substance is also required. This is a living substance called a "ferment."

What is Yeast? Yeast is a ferment. If a small drop of it be placed under the microscope, a large number of minute bodies will be seen. It is these little bodies that excite a fermentation in warm mixtures when sugar is present; this is called " vinous fermentation."

What is Beer? Water is added to the barley until it sprouts. when the starch is changed into sugar. Then heat is applied to kill the young sprouts and to drive off the water. The barley is now called " malt." This malt is ground, and then soaked in water, in order to dissolve all the sugar. The sweet liquid is preserved, and yeast added to it, which causes vinous fermentation; and this, as we already know, changes the sugar into carbonic-acid gas and alcohol. The carbonic-acid gas rises to the top and escapes in bubbles, while the alcohol remains mixed with the water. The result is a poisonous liquor, either beer, ale, or porter.

Yeast in Bread. But yeast is also used in making bread; so what is the difference between putting barley, water, and yeast together and getting beer, and putting wheat flour, water, and yeast together and getting bread? In making bread, yeast is added to the moistened flour. This yeast acts upon the small amount of sugar present, changing it into carbonic-acid gas and alcohol. The gas becomes imprisoned in the sticky dough, making a great number of larger or smaller openings, which cause the bread to be light and porous. As the bread is baked the heat of the oven causes the small amount of alcohol to evaporate and pass out of the bread, together with the gas.

What is the Difference?

In making Bread.	*In making Beer.*
1. Starch.	1. Starch.
2. Sugar.	2. Sugar.
3. Yeast.	3. Yeast.
4. Alcohol and carbonic-acid gas.	4. Alcohol and carbonic-acid gas.
5. The alcohol *is evaporated.*	5. The alcohol *is retained.*

Result.

A valuable food, free from any poison.

A drink, containing a powerful poison.

The Amount of Alcohol in Fermented Liquors.

In 100 parts of cider there are from 5 to 7 parts of alcohol.
"   "   beer   "   "   5 to 7   "   "
"   "   sherry wine   "   15 to 20   "   "

What is Cider? Cider is a drink made from the juice of apples. When this juice is first pressed from the fruit, there is no alcohol in it; but if it be exposed to warm air, it will begin to ferment in from six to eight hours. It is not necessary to add yeast to cause fermentation, for the ferments are always present on the surface of ripening fruits. Fermentation is often hastened because there is mixed with the freshly pressed juice a little of the juice remaining in the mill from a previous grinding. This juice may already contain alcohol; and thus it follows that it is very difficult indeed to get any cider, even fresh from the mill, that does not contain some slight amount of alcohol. When bubbles begin to rise and the froth gathers, it is a sign that the sugar is turning to alcohol.

We have learned that one drink containing alcohol may create a desire for another, and there is great danger that the use of cider, with its alcohol, will create a desire for drinks containing more alcohol; until from this simple beginning, an appetite for the strongest drinks may be formed.

Wine? The ferments that change sugar to alcohol are found on the surfaces of fruits, but never inside. If the juices of such fruits as grapes and currants are pressed out the ferments are washed into them and change their sugar to gas and alcohol. Such juices are called wines. The alcohol remaining in the wine makes it poisonous. A wine may be home-made, and free from any poisonous drugs that might be added to give it color or taste; yet it contains the poison alcohol, and is on this account alone positively injurious.

Acetous Fermentation. Wine and other such drinks are the result of vinous fermentation; besides this there is another fermentation, known as acetic, or acetous, fermentation.

Causes of Decay. In addition to the ferments that produce alcohol there are other minute living forms that cause other forms of decay. One kind belonging to the class called "bacteria" causes acetic fermentation. Other kinds of bacteria cause meat to putrefy and milk to sour. Other minute forms, not bacteria, cause bread to mould and fruit to rot.

How to prevent Decay. If we keep any article of food free from these minute living forms, it will not decay. It may become dry from the evaporation of its water, but it will never decay. These minute forms may be killed by boiling, and they do not grow rapidly in either a cold or a hot mixture. Milk keeps better in a cool place simply because they do not thrive so well where it is cool. Scalding milk makes it keep longer, because the heat checks the growth of these minute bodies.

The Object of canning Fruit. These small bodies are so light that they easily float in the air, although they are so minute that they cannot be seen; we can see only their effects. The object of canning fruit is to keep it from their destructive work. The heat used in canning the fruit kills those that are present, and the tightly sealed cans prevent others from entering. If the top of the can be loosened, and fresh air be allowed to enter, then some of these small bodies will surely enter also, as they are always present in the air. It is these, and not the air, that make the fruit ferment, or sour.

Dust in the Air. That vast numbers of particles of matter float in the air, any one can decide for himself. It is necessary only to look at a ray of light as it enters a darkened room, to see in it vast numbers of dust particles. The minute bodies that cause decay are so much smaller that it would take many thousands of them to make a mass as large as the smallest particle of dust.

2

Is there Alcohol in Vinegar? When the juice of the apple, or other fruit, has undergone vinous fermentation, alcohol is formed in it. If this be allowed to remain in the vegetable juice, and be kept warm by exposure to the sun, or in some other way, another kind of ferment will enter the fluid and change the alcohol into a sharp acid, called acetic acid, and commonly known as vinegar. There is no alcohol in vinegar. It is used to flavor food, and is another illustration of how a ferment changes the substance it works upon.

QUESTIONS.

1. What are the two classes of liquors?
2. How are corn and barley treated to make liquor?
3. What things are necessary to produce alcohol?
4. Tell something about yeast.
5. Describe how beer is made and its character.
6. Describe how bread is made.
7. What becomes of the alcohol in each?
8. Is there any alcohol in the juice of apples before it has fermented?
9. Why not drink cider?
10. Are alcoholic ferments found inside the fruits? How then can these ferments make wine? What is the character of wine, and why?
11. What causes acetous fermention?
12. How can we prevent food from decaying? Why do we can fruits?
13. Tell something about vinegar. Does it contain alcohol?

CHAPTER III.

DISTILLATION.

WE have spoken of alcohol after it is obtained by fermentation and is still mixed with the fruit or plant juices. We must now describe how it is obtained in a pure state from these mixtures, and also how the various distilled liquors are made.

What is Distillation? A very easy experiment will show the principle of distillation. Notice the steam coming from a kettle of boiling water. This is water changed to vapor by heat. Hold a cold cup over this steam, and notice the drops of water gathering on it. The cold cup has condensed the vapor to a liquid again. This is called "distillation," because that word means "to flow in drops." Distillation, therefore, is simply changing a liquid to vapor, and then changing the vapor back to a liquid again.

Alcohol evaporates easily. Alcohol evaporates very readily when exposed to the air, and will rapidly pass into vapor if heat be applied to it. It will evaporate with much less heat than is required for water to evaporate. That is, if we have a mixture of alcohol and water, we can apply a very little heat and cause the alcohol to pass off in a vapor, before the water will be warm enough to boil.

The Way Alcohol is obtained. Now, if we take fermented liquors, which contain alcohol, water, and other ingredients, and expose them to a low heat, we can drive off the alcohol as a vapor, and leave the other ingredients behind. All that remains to be done is to condense this vapor, and we have alcohol.

Brandy, Whiskey, and Rum. When these liquors are made, there is just enough heat applied to the fermented liquid to cause not only the alcohol to pass off, but also some of the water and certain other substances. When these vapors are all condensed, they make brandy, whiskey, or rum. Each of these consists of at least one half clear alcohol, and sometimes they are even much stronger. Brandy is distilled from wine and cider; whiskey from fermented corn, barley, or other grains; rum from fermented molasses. Sometimes these liquors are distilled two or three times, in order to make them stronger.

Pure Liquors. There is a popular impression that "pure" liquors are not harmful, and that if prepared in a careful manner they are even useful. This is surely a false impression; for the purest liquors contain alcohol, and it is the presence of that poison which works such injury. Bad, however, as the purest of them are, they are often made more poisonous by having mixed with them various coloring matters and poisonous drugs. Distilled liquors cause speedier destruction than the fermented because of the larger amount of alcohol they contain, but the use of the fermented leads to a taste for the distilled.

CHAPTER IV.

CELLS.

Anatomy treats of the structure of the body.

Physiology treats of the uses of the body.

Hygiene treats of the methods by which health is preserved.

To illustrate these definitions, we say that Anatomy describes the stomach ; Physiology tells what it does ; and Hygiene teaches us how to keep it healthy.

We learn from Anatomy that the body is composed of bones, muscles, nerves, and blood vessels. From Physiology we learn what all these parts are for, — how the blood circulates ; how we breathe ; and how we digest our food. Hygiene teaches us that there are certain laws of health, and that we must obey those laws or suffer the penalty. It tells us what kinds of food we should eat, and informs us as to the care we should take of the whole body.

The Eye requires Aid. The eye is a most wonderful organ, yet there is a limit to its capability. The telescope and the microscope are used as aids to the eye. By means of the telescope the astronomer can see many stars that are entirely invisible to the naked eye. The microscope shows many things about the body that could not be seen without it. The botanist uses it in his work, and he finds by it that every flower and fruit, every blade of grass, and every piece of wood is made up of the most minute parts, called " cells."

A Potato. To illustrate this, let us examine the inside of a potato. With the unaided eye it looks as if it consisted of one common substance. But the microscope shows that it consists of large cells ; and in these cells are vast numbers of little grains containing starch, which are hence called starch grains.

How to obtain this Starch. Cut a potato in thin slices, and place these in water. By stirring them you can make the water look milky. Now remove the pieces of potato, and let the water become quiet. A white powder will settle to the bottom of the dish. This is pure potato starch, and when placed under the microscope will show some oval bodies, as illustrated in Fig. 1.

Other Starches. We take the wheat to the miller in order that he may remove some layers of cells that are indigestible, and return the rest to us as wheat flour. The starch grains in this do not look like those of potato ; they are smaller, and more nearly round. (Fig. 1.) Oat starch that is used for cooking is composed of grains shaped otherwise, as is seen in the figure. Even a leaf is shown by the microscope to consist of minute cells.

Cells in the Vegetable Kingdom. Thus we learn that every part of the vegetable kingdom is composed of cells.

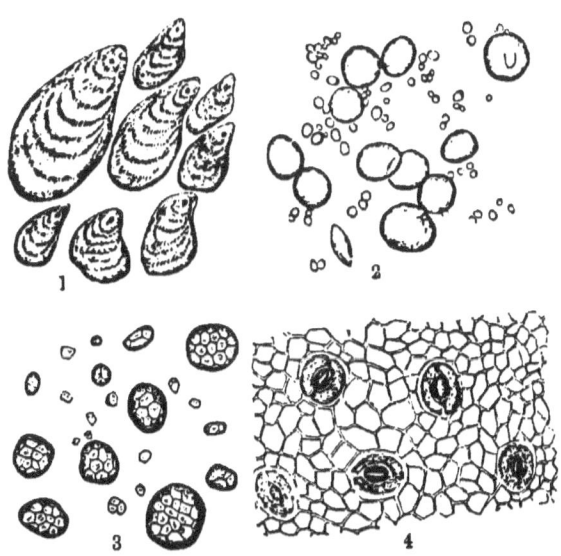

FIG. 1. Starch grains: (1) from the potato, — potato starch grains; (2) from wheat, — wheat starch grains; (3) from oat, — oat starch grains; (4) cells from the surface of a leaf.

Cells in the Animal Kingdom. The same thing is true of the animal kingdom. When we look at a drop of blood we little think that the microscope would show in it vast numbers of minute cells; and yet there are as many as five millions of them in every drop. Then we look at the skin, and think it is one solid mass of covering; but the microscope shows that it is made up of a number of layers of cells.

The Whole Body a Collection of Cells. By means of the microscope we learn that all the organs and tissues of the system are composed of minute bodies, called cells. (Fig. 2.)

The Cells are not alike. There are round cells, long, narrow cells, and cells of all shapes and sizes. Some of

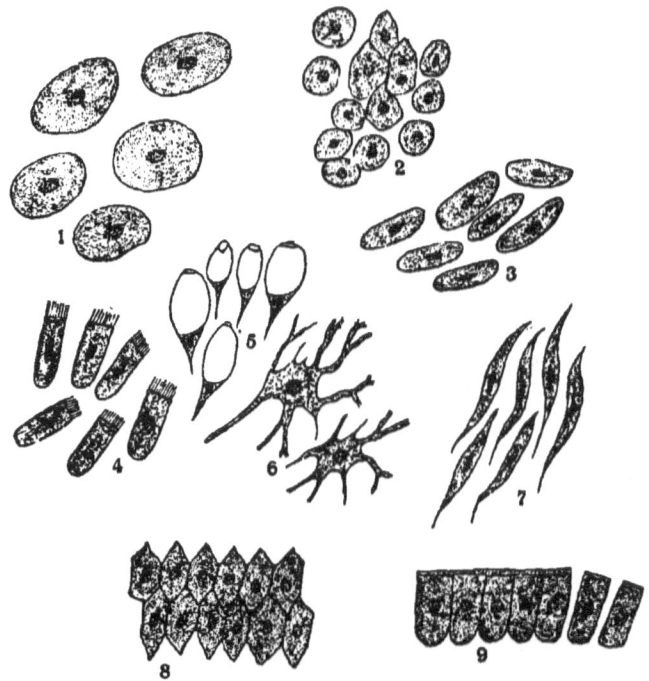

FIG. 2. Cells from different parts of the body, —(1) from the inside of the cheeks; (2) from the liver; (3) from the nail of a finger; (4) from the bronchial tubes; (5) from the intestines; (6) from beneath the skin; (7) from muscle; (8) from the eye; (9) from the stomach.

them are so small that it would take three or four thousand of them, side by side, to make an inch in length;

while others are so large that they can almost be seen with the unaided eye. Some are colorless, others are lightly colored, and still others are jet black.

What is Histology? After we have studied the body as far as we are able with the unaided eye, there is still much to be learned by means of the microscope. Microscopic anatomy, or histology, is a description of the minute structure of animals or plants.

QUESTIONS FOR CHAPTER III.

1. Illustrate the principle of distillation.
2. What is distillation?
3. Which evaporates more readily, alcohol or water?
4. How is alcohol obtained from fermented liquors?
5. How much clear alcohol is there in brandy and whiskey?
6. Are " pure " liquors harmful?
7. What is in the purest liquors that makes them so injurious?
8. Why are distilled liquors worse than fermented ones?
9. What leads to a taste for the distilled liquors?

QUESTIONS FOR CHAPTER IV.

1. Of what do anatomy, physiology, and hygiene treat?
2. Illustrate, or explain each.
3. Of what use is the microscope?
4. What would the microscope show in a potato?
5. Are any cells found in the vegetable kingdom?
6. Are any found in the animal kingdom?
7. What does the microscope show in the blood and the skin?
8. The whole body is a collection of what?
9. Tell what is said about cells.

CHAPTER V.

THE BONES.

The Frame-work. In building a house it is necessary that there should be a strong frame-work; for on this will be fastened the boards, and on the boards will be placed the shingles, and the whole will be beautifully painted. The frame-work of a house is so constructed that there are many rooms in it. These rooms have openings between them, so that things can be carried from one part of the house to another. There is a frame-work of our bodies, dividing each body into a number of rooms, and giving a firm support, to which many important parts can be attached. This frame-work also serves as a protection to many delicate organs that are placed in some of its rooms; it is composed of bone.

General Description. There are over two hundred bones in the body. Some of them are long, large, and round, while others are thin and flat; while there are others that are so irregular in shape that it is very difficult to describe them. The largest bone in the body is the femur, or thigh bone. (Fig. 3.)

Uses of the Bones. Some of the bones are used principally to protect delicate organs that would otherwise be too much exposed to injury. The eye is well protected in this way. It has walls of bone all about it, except in front, where it is necessary for the light to enter. The brain is completely surrounded by a bony wall, and the heart and lungs are well protected by the bony walls of the chest. Some bones are used for the attachment of muscles, while others are for the purpose of giving proper shape to the body.

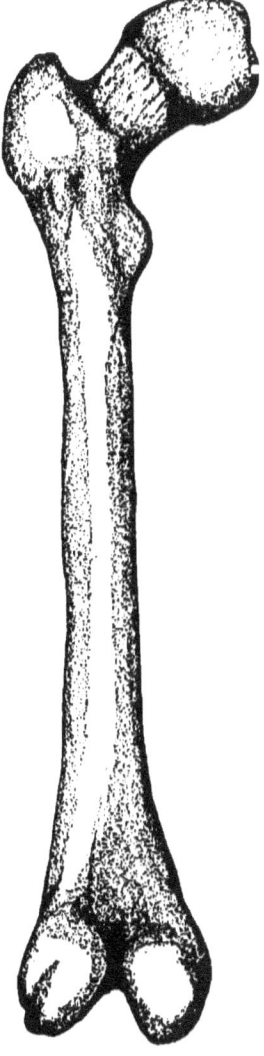

Are the Bones solid? This is easily answered by examining one of the large bones of any animal after it has been sawed open. If it be sawed lengthwise, we shall find that it is hollow, except at the ends. Why is this? It is necessary that the bones should be not only strong, but also light; and it has been found that a hollow bone of a certain length and a certain thickness is both stronger and lighter than it would be if the same amount of material had been solid and of the same length.

The ends of the long bones are filled with little bands of bone, giv-

Fig. 3. The human femur, or thigh bone.

ing them a honey-combed or spongy appearance. If one of the thin, flat bones of the head, or one of the small bones of the hands or feet, should be cut open, there would be found in it this same spongy bone.

The Marrow of Bone. The large cavity inside the long bones, and all the spaces in spongy bone, are filled with a yellowish or reddish substance, called "marrow." This is composed of fat and cells.

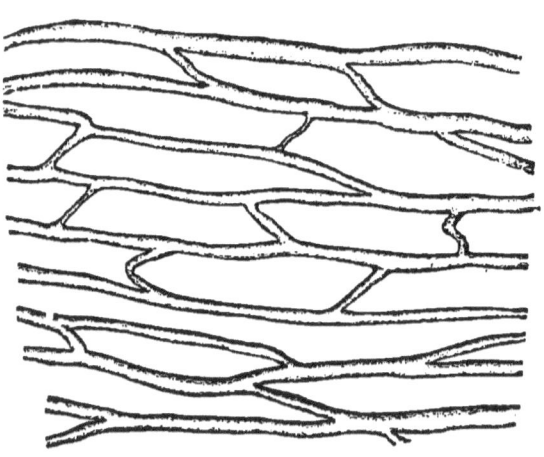

FIG. 4. The blood vessels of bone, as seen with a microscope.

Can Bone bleed? The bones are well supplied with blood vessels, even the hardest part of a bone having in it an immense number of small ones. (Fig. 4.) Nearly all these blood vessels are so small that they can be seen only by means of a microscope; yet if we examine the outside of a large bone of any animal, we can usually find one or more holes through which a blood vessel has passed.

The Soft Parts of Bone. Besides the blood vessels, there are countless numbers of soft cells all through the bones. (Fig. 5.) All the black, spider-shaped bodies in the illustration are the cells of bone as they appear when examined with a microscope. The soft parts of bone, therefore, are composed of blood vessels, cells, and

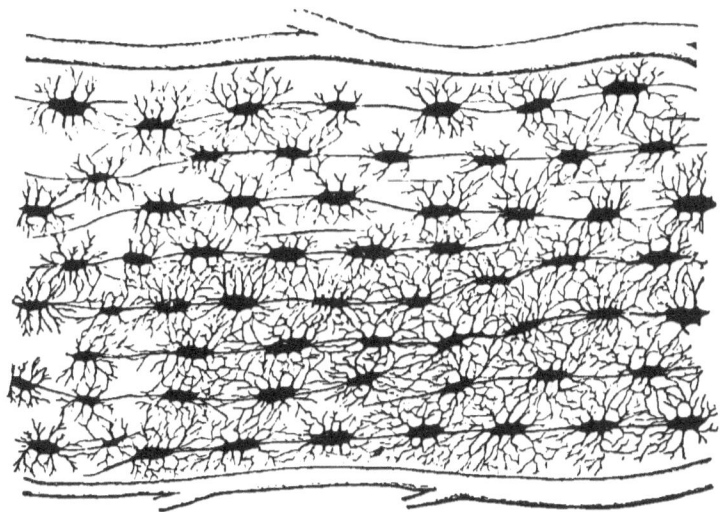

FIG. 5. Human bone, showing two blood vessels and a number of bone cells, magnified.

nerves, and a frame-work of soft fibres. These are called the animal matter of bone.

The Animal Matter. The animal matter can all be taken out of a bone by simply putting it into the fire. The shape of the bone will not be changed, it will only become lighter and whiter. After the animal matter has all been burned out, the bone can easily be broken, and pounded into a fine powder. On the other hand, if the bones were composed of animal matter alone, we could bend them into any shape without their breaking; in fact, they would be of little use to us in this condition.

The Mineral Matter. The hard part of bone is composed largely of lime, which may be removed by soaking the bone in a weak acid for a few hours. The bone will retain its former shape; but it can be easily bent, and even tied in a knot, as shown in the illustration. (Fig. 6.)

FIG. 6.   A bone tied in a knot, after the mineral matter has been removed by an acid.

The blood vessels and the cells are still in this bone, while the soft fibres of tissue keep it in shape.

A very interesting experiment can be made to illustrate the action of an acid on mineral matter. Place an egg in strong vinegar for a few hours. The acid of the vinegar will dissolve the lime, and leave nothing but a soft membrane for a covering. The egg can now be pushed through the opening of a bottle much smaller than its own diameter. As soon as it is in the bottle it will assume its original shape; and the puzzle is to have your friends discover how such a large egg could be put through so small an opening.

Bend, or break? From what has been said, it is very clear that if the bones do not contain the proper amount of mineral matter, they will bend before they will break. On the other hand, if they do not contain enough animal matter, they will easily break.

In early life there is always an excess of animal matter; hence children will tumble about and have heavy falls without breaking a bone. Sometimes children are allowed to walk before there is enough mineral matter in their bones to bear much pressure; then the weight of the body causes the bones of the legs to bend. We say such children are "bow-legged."

Young Bones may have their Shape changed. When the bones have an abundance of animal matter in them, as in early life, they can be moulded, and their natural form greatly changed. The Chinese know this, and so they bind the feet of their young children until the bones are so changed that one can hardly recognize them as belonging to the feet. (Fig. 7.) Some tribes of Indians bind the heads of their children to boards in order to change their shape.

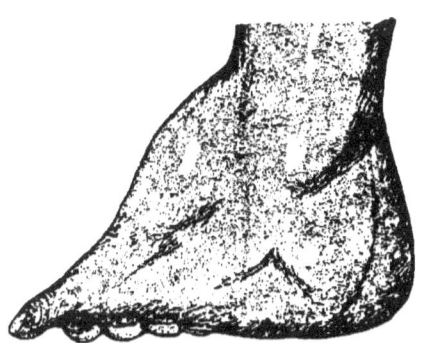

FIG. 7. A Chinese foot, deformed by bandaging. (From a plaster cast.)

Nature changes the Shape of Bones. In order that some of the bones may better perform their work, Nature causes their shape to be changed. (Fig. 8.) This is well illustrated in the case of the lower jaw, which changes its shape as the person advances in years.

The Covering of Bone. Surrounding each bone is a thin membrane, called the "periosteum," which is essential to the life of the bone. It is well supplied with

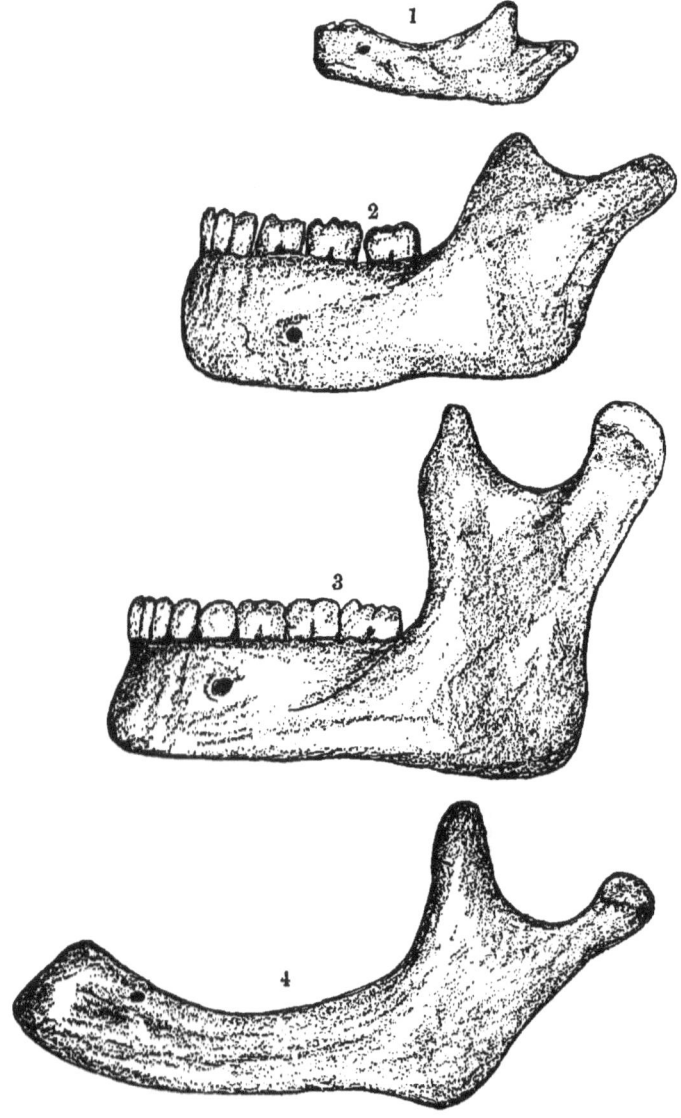

FIG. 8. (1) is a side view of the lower jaw at birth; (2) at fifteen years of age; (3) at thirty years of age; (4) at seventy years of age.

blood vessels, some of which pass directly into the bone through the canals which we have already mentioned.

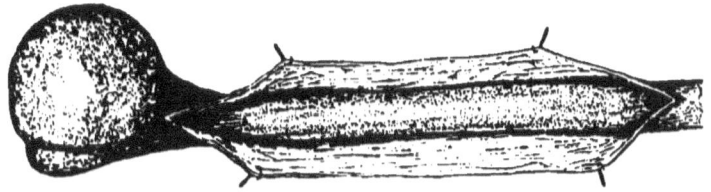

FIG. 9. A bone with the periosteum partly peeled off.

The illustration shows how it is possible to peel off this membrane from any fresh bone.

QUESTIONS.

1. What is the object of the framework of a house?
2. What is the object of the bony framework of the body?
3. Give a general description of the bones.
4. What are some of the uses of the bones?
5. Tell about the protection of the eye, the brain, and the heart.
6. If a large bone be sawed lengthwise what will it show?
7. Where is spongy bone found?
8. Describe the marrow of bone.
9. Describe the blood vessels of bone.
10. Tell something about the soft parts of bone.
11. What is meant by the animal matter of bone?
12. How can you get rid of it?
13. Why not have the bones all animal matter?
14. Of what is the hard part of bone composed?
15. How can you get rid of the mineral matter?
16. Can the shape of bones be changed?
17. What is the periosteum?

CHAPTER VI.

THE SKELETON AND THE JOINTS.

The Bones of the Skull. The bones of the skull make a complete covering, or tight box, for the brain, with only a few holes in it to allow the nerves and blood vessels to pass in and out. These bones also protect the organs of sight, smell, hearing, and taste.

The Bones of the Trunk. The bones of the trunk are the spinal column, or backbone, the ribs, and the breast-bone.

The Spinal Column. The spinal column consists of twenty-four small bones, resembling the one in the figure, and two irregular ones at the lower end of these. Between these bones, and attached to them, are soft cushions of gristle or cartilage. These cushions act as springs, so that running or jumping, or even walking, may not jar the body too greatly. (Fig. 10.) If all these cushions could be piled up together, they would make a mass over six inches in thickness, and it would be as elastic as so much rubber. If we walk or stand very much during the day, these cushions become flattened; but during the night they regain their former thickness. On account of this elasticity we are a trifle shorter at night than in the morning. For the same reason we are a trifle shorter when standing than when lying. If it were not for these cushions, we could not run or jump, or even walk, without jarring, or perhaps greatly injuring, either the brain or the spinal cord.

Each one of the bones of the spinal column is called a "vertebra." Each vertebra has a large opening in it (Fig. 11.), so arranged that when all these bones are properly put together, the openings will form a canal, in which the spinal cord rests. This canal is continuous with an opening at the base of the skull, so that the spinal cord and the brain may be connected.

The illustration (Fig. 10) indicates that the spinal column is not straight. It is bent backward near the shoulders, and forward near the waist.

The Ribs. There are twelve of these slender, curved bones on each

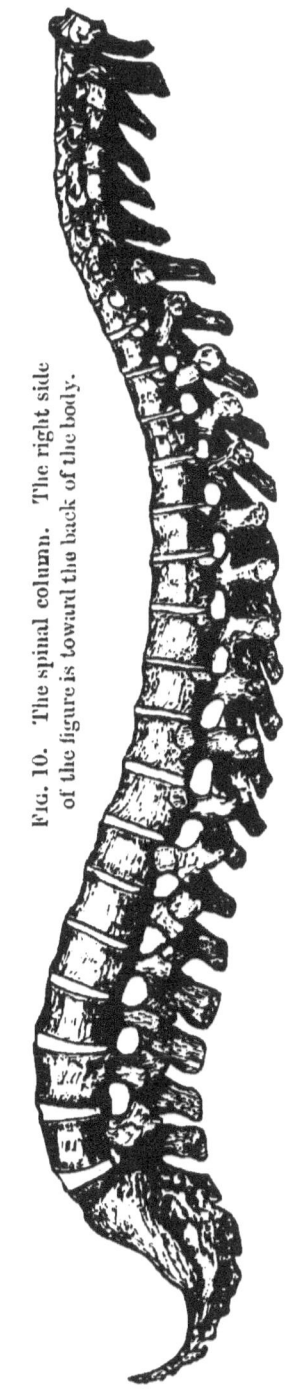

FIG. 10. The spinal column. The right side of the figure is toward the back of the body.

side of the chest. Behind, they are attached to the spinal column, while in front, some of them are attached to the breast-bone, three to each other, and two are not joined to anything; these two are called the "floating ribs."

If we look at the figure (Fig. 45, p. 131) of the chest, or thorax, we can understand how easy it would be to bring the ends of the lower ribs nearer together, by tying a cord or band around the lower part of the chest. Some persons seem to think that they know better than Nature what the shape of their chests should be; so they try to make their waists as small as they can. They call it "fashion;" others call it "tight lacing;" while the doctors call it "a great wrong."

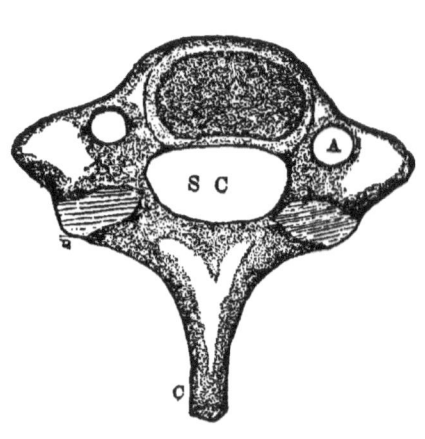

Fig. 11. One of the bones of the spinal column. A is an opening in each side for a blood vessel; B is the point on which the bone above it rests; c is the long process that extends from the back of the spinal column; s c is the large opening in the centre for the spinal cord.

The Upper Limbs. There are five large bones and several small ones that belong to the upper extremity.

The collar bone extends from the front of the shoulder to the top of the breast-bone.

The shoulder blade is behind the arm and between it

and the spine. The outer ends of these two bones and the upper end of the arm make the shoulder.

From the shoulder to the elbow there is one large bone, called the "humerus." From the elbow to the hand, which is known as the "fore-arm," there are two bones.

A number of small bones form the wrist and the hand, and these are so arranged as to allow a variety of movements.

The Lower Limbs. The bones of the lower extremity of the body are very like those of the upper.

The pelvic bones can be felt on the sides of the lower part of the body. These, with the upper end of the thigh bone, form the hip. From the hip to the knee there is one large bone, known as the "femur." From the knee to the foot there are two bones. A small bone, called the "knee pan," covers the front of the knee joint.

There are a number of small bones in the foot, arranged, as in the case of the wrist and the hand, so as to render a variety of movements possible.

The Arch of the Foot. The bones of the foot are so put together that they form an arch, so that only the front and the back of it touch the floor. This arch is useful in protecting the body from too severe a shock when one is running or jumping ; for when the weight of the body is thrown upon the arch, its centre is pressed downward, and thus it acts like a spring.

The Joints. There may be only two, or there may be many bones forming a joint.

The Joints are oiled. On the ends of the bones which come together to make a joint there is a kind of gristle, called " cartilage." This is covered with a very thin membrane, which is constantly secreting, or pouring out, a watery substance. This substance, called the "joint water," serves the same purpose that oil does when it is put upon the joints or

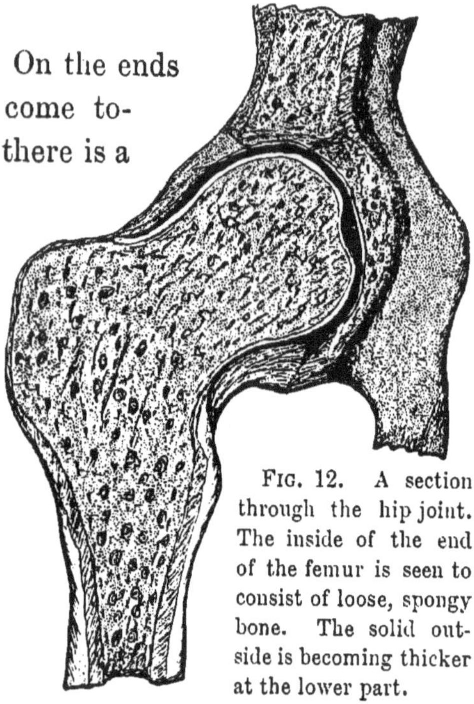

FIG. 12. A section through the hip joint. The inside of the end of the femur is seen to consist of loose, spongy bone. The solid outside is becoming thicker at the lower part.

wheels of machinery. If the membrane covering the cartilage should become inflamed or injured in any way, it might fail to furnish enough of this fluid; then the movements of the joint would be painful, and the joints themselves would become stiff and misshapen.

The Ligaments. The bones are held in place at the joint by means of white, shining bands of tissue called " ligaments." A " sprain " is an injury to the ligaments.

Tight and Loose Joints. The ligaments of some persons are very firm, and the joints are not easily moved. In others the ligaments are not so firm, and the joints are more easily moved. We say of these latter that they are " loose jointed." When a bone gets " out of joint," it breaks its way through these ligaments.

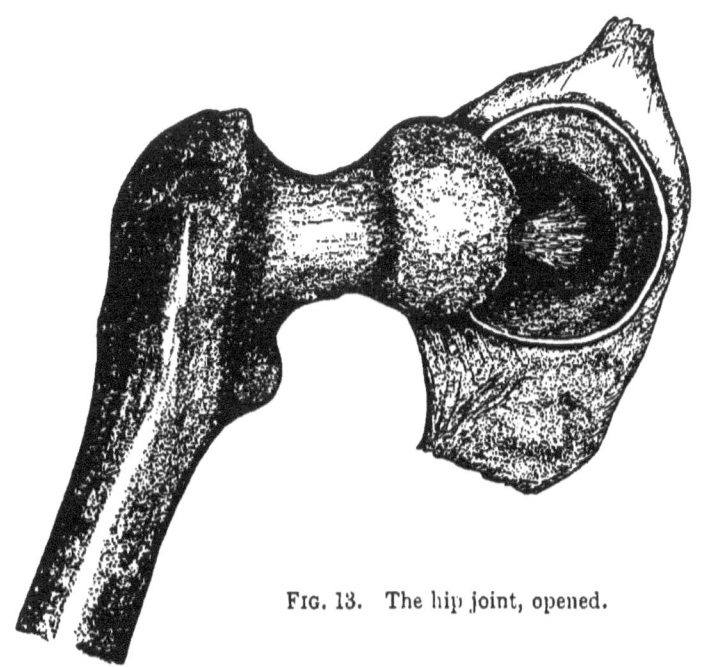

FIG. 13. The hip joint, opened.

Different Kinds of Joints. Most of the joints are either " ball-and-socket " joints, or " hinge " joints. The hip and the shoulder are good illustrations of the former, for they allow movements in every direction. The elbow is an illustration of a " hinge " joint, because. like the hinge of a door, it allows movements only in two directions.

Fig. 12 shows the ball resting tightly in the socket. Fig. 13 shows the hip joint, with the outer ligaments removed. Only one ligament is left, — that which is seen to connect the centre of the ball with the centre of the socket.

QUESTIONS.

1. What do the bones of the skull protect?
2. Name the bones of the trunk.
3. How many bones in the spinal column?
4. What is between these bones?
5. What are these cushions for?
6. What is a vertebra?
7. What is its large opening for?
8. How many ribs are there?
9. How many large bones in each upper limb?
10. Where is the collar bone?
11. Where is the shoulder blade?
12. Where is the humerus?
13. How many bones in the forearm
14. Where are the pelvic bones?
15. Where is the femur?
16. What bone covers the front of the knee joint?
17. How is the arch of the foot useful to us?
18. What covers the ends of the bones forming a joint?
19. What keeps the joints oiled?
20. What are the ligaments?
21. What is a sprain?
22. What happens when a bone gets out of joint?
23. Give some illustrations of the different kinds of joints.

CHAPTER VII.

HYGIENE OF THE BONES.

We have already learned that the bones of young persons are easily bent because there is so much animal matter in them. So it is possible, simply by improper ways of walking and sitting, to deform the body and make it crooked and irregular.

How to have a Good Form. If we wish to have a fine, erect form, we must endeavor to keep the bones of the spine in their natural position. If we do this when we are young, and when the bones are daily becoming more solid, then when our bodies are well developed we shall find that we cannot with comfort walk or sit crooked.

How to walk. To keep the spine in its natural position we should walk with the shoulders thrown well back and the whole body erect. If a person gets in the habit of " stooping " when he walks, or of sitting at the table " all bent over," the elastic cushions between the vertebræ will become so compressed on one side that they will lose their elasticity there, so that when he tries to straighten up, it is of no use; he is "round-shouldered," and will remain so.

Straight, or Crooked. Let us look at two persons. One stands as "straight as an arrow;" he sits upright at the table, and people say, "What a fine form!" We should not pick out such a person as likely to have consumption, for "he looks too healthy."

The other person is round-shouldered, his chest is narrow, it is an effort for him to run or jump, and he is subject to coughs and colds. We wish to tell him to "straighten up," but it would do no good for him to try now; the bones are well filled with mineral matter, and the elastic cushions are old and unyielding. In youth we should throw off any "tired feeling" we may have, and keep erect, that we may make sure of having a good form and a healthy body for the future.

How to Stand. When some persons stand, they rest their weight on one foot; this habit is sure to make the hip bones grow out of shape. It will bend the spine, and, sooner or later, make it incline toward one side. As a rule, it is better to stand with the weight of the body on both feet. To do this, one thing is certainly necessary, — there must be easy shoes to stand in.

Proper Shoes. When buying shoes it is customary to have them fitted while seated on a chair or couch, and the dealer is cautioned "not to get them too large." So a good, snug fit is made. The arch of the foot is forgotten; and it follows that when the weight of the body is thrown on the feet, the arch is flattened and the foot lengthened. The shoe is now altogether too small, and great discomfort and harm result. These consequences will be all the more marked if the shoe be made with a high heel. A high heel throws the whole body out of line, and is the cause of a number of most distressing complaints. The low " common sense" heel is far better, and we hope our young friends will always choose it, showing, by their preference, that they care more for a " healthy body " than for fashion.

Support for the Feet. There ought to be a support for the feet in all schoolrooms ; the seats should be low enough to let them rest on the floor.

Form, not deform. The bones are not fully developed until the person is at least twenty-five years of age ; and it should be remembered that even after this they in some degree change their soft substance. We ought to do everything in our power to aid in forming our bones, rather than in deforming them.

ALCOHOL AND THE BONES.

If we remember that the bones are filled with blood-vessels, nerves, and cells, and that these cannot be in a sound condition unless the whole body be healthy ; and if we remember that all forms of alcohol seriously affect the animal matter of the bone while it is in a growing state, — we may appreciate the effect on the body of that drug. By acting through the nerves and blood, alcohol has great power in retarding the development of the frame-work of the body.

The early use of tobacco also will seriously affect the proper development of the bones. It may be put down as a very general rule that the early use of tobacco and alcoholic drinks is likely to stunt the growth of the bones, and thus actually to dwarf the whole body.

QUESTIONS.

1. Why are the bones of young people so easily bent?
2. What bad effects may result from this fact?
3. How should we walk?
4. What may happen to the elastic cushions by improper methods of walking or sitting?
5. Should the whole weight be thrown upon one foot while standing?
6. What is said about fitting shoes?
7. What is said about the heels of shoes?
8. Does alcohol have any effect on the growth of bones?
9. Does tobacco affect their growth?
10. What is said about the early use of tobacco and alcoholic drinks?

SUGGESTIONS TO TEACHERS.

1. Procure from the meat-market a piece of fresh bone that has been cut transversely. Observe the oily matter on the inside; this soft material is the **marrow**.

2. Take this same bone and cut down upon it with a knife, and thus peel off the covering membrane. This is the **periosteum**.

3. Procure any bone and burn it in the stove. Show how easily this can be broken by striking it with a hammer. Show how easily it can be pulverized by a continued use of the hammer. Try hammering on a fresh bone, and call attention to the marked difference in the results. This experiment demonstrates the existence of **mineral matter** in bone.

4. Procure one of the bones of a fowl, such as the leg bone of a chicken, or even one of the ribs. Place it in a solution consisting of water three parts, and nitric acid one part. After it has been in the acid a few days, place in water for a short time, in order to wash out the acid. The length of time required for the acid to complete its work will depend on the size of the bone. For a small bone three or four days will suffice; for a larger bone, a week may be required. The acid has done its work when a needle can be pushed through the bone. The bone can now be bent and doubled, and even tied in a knot. What remains after the elimination of the mineral matter by means of the acid is the **animal matter**.

5. Take any old dry bone. Saw it in two lengthwise. Observe that the bone is hollow except near the ends, which are filled with a honey-combed arrangement of bone. This is the **spongy bone**.

6. Look on the surface of any dried bone for minute openings. When found, they can be shown as representing the places where the larger **blood vessels enter the bone.**

7. It is easy to procure at the meat market the smooth, round end of the hip bone of some animal. Have this cut lengthwise. When the cut parts are placed together, they will show the **ball** of the " ball-and-socket " joint. In the centre of the end the **ligament** is attached, as illustrated in Fig. 13. When the cut parts are separated they will show the layer of **cartilage** that covers the end of the bone.

8. The **elastic cushion** between the vertebræ may be procured in any meat market, as well as the two adjoining vertebræ, cut lengthwise. The relative amount of cartilage is readily comprehended, as well as its use.

CHAPTER VIII.

THE MUSCLES.

Their Number, Size, and Purpose. There are more than five hundred muscles in the body, nearly all of which are arranged in pairs, so that the two sides of the body are almost alike. Some of these muscles are very small, while others reach from the hip to the knee. Their primary object is to move the different parts of the body; but they also aid in giving proper shape to the body, and in enclosing cavities, — as the mouth. Nearly half the weight of the body is due to muscle.

Two Parts of a Muscle. The muscles that are under the control of the will consist of two parts, — the large red portion, called the body; and the white, shining ends, called the tendons.

The Tendons. The tendons are easily seen by examining the muscles, or flesh, on the leg of a fowl, after the removal of the skin. They can also be felt at the wrist when the fingers are moved.

Nearly all the muscles of the fore-arm terminate in tendons near the wrist. The tendons are fastened to the fingers, so that when the muscles of the arm or fore-arm contract, they draw these tendons, and thus move the fingers.

Size of the Tendons. The tendons are always much smaller than the muscles to which they belong. This

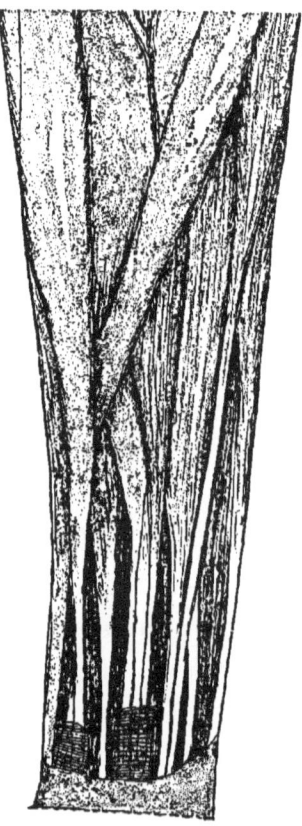

is well illustrated in Fig. 14. The muscles at the middle of the fore-arm, as shown in the figure, are quite large, while the tendons at the wrist are so small that it can be spanned with the thumb and finger. This makes the movements at the wrist more free and easy.

The largest tendon in the body is at the back of the leg. It is attached by a muscle to the heel, so that when that muscle contracts, the tendon draws up the heel, as in walking.

The muscle is the active part; the tendon is only a cord which can be pulled by the muscle.

FIG. 14. The muscles of the arm, ending in the white tendons at the wrist.

Two Kinds of Muscle. The muscles are divided into two classes, — the voluntary, and the involuntary.

We can move some muscles whenever we wish, as those of the face and the arm. Because we are thus able to control their movements, they are called voluntary. But some muscles cannot be controlled in this way. They do their work whether we wish it or not. We cannot control their movements by the will, so they are called involuntary. The muscles of the stomach and the heart are of this variety. The heart beats and the stomach contracts, and we have no power to stop them.

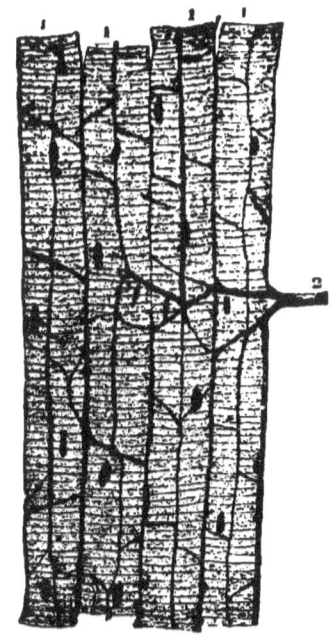

FIG. 15. Voluntary muscle, with its blood vessels. (1) the muscle fibres; (2) the blood-vessels, magnified.

The Uses of Muscle. Nearly all the voluntary muscles are attached to bone at each end; it is because they contract and move the bones that we are able to run and jump and perform all the movements of which the body is capable.

The Structure of Muscle. If a piece of boiled, lean meat, which is voluntary muscle, be examined, it will be noticed that it readily falls apart into little threads of tissue; and with needles these little threads may be easily divided into still smaller threads. If one

4

of these small threads be placed under the microscope it will be found to consist of many small fibres. (Fig. 15.) Therefore we say that voluntary muscle is made up of many small fibres, all bound together. Our illustration shows four of these fibres, and also the blood-vessels that belong to them.

Involuntary muscle is composed of very small cells which are placed close together. The illustration shows them separated from each other, so that their shape may be better seen. (Fig. 16.)

FIG. 16. The cells of involuntary muscle, magnified.

The Contraction of Muscle. The muscles are of use to us because they have the power to contract.

When a muscle contracts, it becomes thicker, harder, and shorter. (Fig. 17.) That it becomes harder and

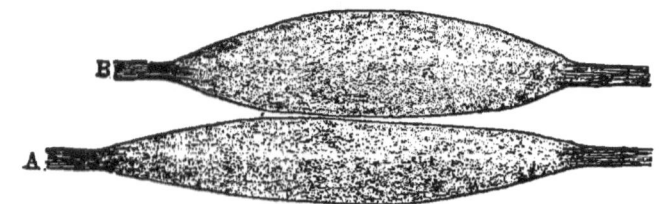

FIG. 17. A, a muscle relaxed, before it contracts ; B, the same muscle contracted. It is shorter and thicker.

thicker, is easily shown by placing the thumb on the end of the little finger, and pressing the fingers of the other hand on the ball of the thumb. It can also be shown by placing the hand on the front of the arm and raising the fore-arm ; the muscle will be felt to swell and harden.

That the muscle shortens, is proved by the fact that it moves the parts to which it is attached. All the movements of the body are made by the contraction of its muscles.

The two diagrams below show how the contraction, or shortening, of the muscles causes the parts to which they

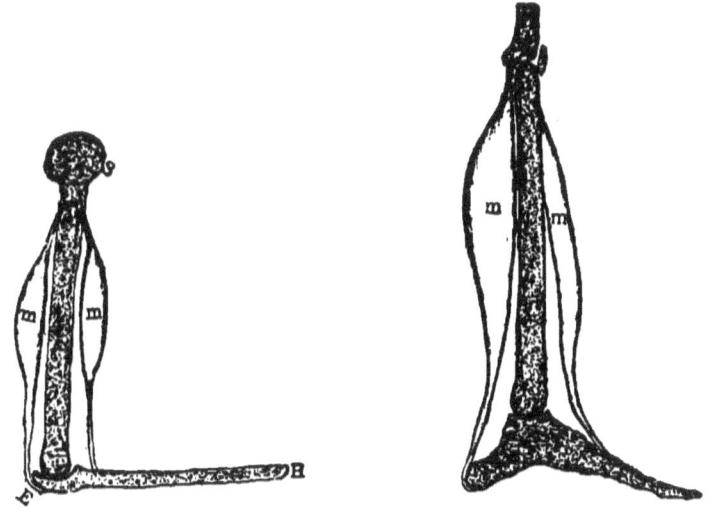

Diagrams illustrating the action of muscles.

FIG. 18. Muscles of the arm: S, the shoulder ; E, the elbow ; H, the hand; m, muscles.

FIG. 19. Muscles of the leg.

are attached to move. In Fig. 18 it is clear that if the muscle on the front of the arm should contract, the fore-arm and the hand would be raised ; while if the muscle on the back of the arm should shorten, they would be drawn down again.

This principle is again illustrated in Fig. 19. It is clear that if the muscle on the front of the leg should shorten, it would pull on its tendon and raise the foot; while if the muscle on the back of the leg should shorten, it would pull on its tendon and raise the heel. These illustrations show the principle on which all the voluntary muscles act.

QUESTIONS.

1. How many muscles are there in the body?
2. How are they arranged?
3. Are they all of the same size?
4. What are the muscles for?
5. Describe the parts of a muscle.
6. Where can tendons be found?
7. What is said about the size of the tendons?
8. Where is the largest tendon?
9. Which is the active part, muscle or tendon?
10. Name the two classes of muscle.
11. Why are they so called?
12. Give illustrations of each kind.
13. How is it that we are able to run and jump?
14. Tell something about the structure of voluntary muscle
15. Is involuntary muscle the same?
16. What happens to a muscle when it contracts?
17. How can you illustrate that it becomes harder and thicker?
18. How do you know that it shortens?

CHAPTER IX.

EXERCISE.

Exercise of the Muscles. If a person should place his arm in a sling and not use it for a few months, it would gradually become smaller and smaller. The arm would become smaller because its muscles would first become soft and flabby, and then, at last, would nearly all disappear. But we know that by use a muscle becomes larger, harder, and stronger. The blacksmith always has large arms, and the arm he uses the most will be the larger.

Exercise is necessary. It is absolutely necessary to exercise the muscles, if we wish to keep them healthy and strong. Exercise makes the blood circulate better; so it follows that when we are exercising our muscles we are also giving a better circulation to the blood in the brain and other organs of the body.

Over-Exercise is bad. But over-exercise is nearly as bad as no exercise. It is not a good practice to play or exercise in any way until one is " all tired out."

Exercise all Parts of the Body. We should not exercise any particular part of the body to the neglect of any other part, but we should endeavor to develop all parts equally well.

No one admires a man who is all muscle and no brain ; neither do we like to see one who is very learned, and yet is constantly suffering from disease. Therefore, if we have been studying all day, a brisk walk in the evening will make us feel refreshed ; while if we have been using the muscles at hard work during the day, reading or studying is the proper exercise.

Expression. The various expressions of the face are caused by the contraction of voluntary muscles ; and as a muscle is strengthened by exercise, so it follows that those muscles of the face which are used the most will become the strongest.

If the muscles we use when we laugh are made to contract a great deal, they will become stronger than their neighbors ; so that even when a person is not thinking of laughing, these muscles will exert an influence : there is consequently a slight expression of laughter left on the face. We say such a person has a pleasant smile all the time. If a person cries a great deal, there will be left a slight expression of crying. We say such a person has a sad face. If a person is in the habit of being cross and sullen, it will leave its effect in the features.

As a result of these observations, we establish the rule that the expression which is most constantly on the face will become permanent. This is the reason why it is possible to tell the disposition of a person by the expression of the countenance.

Do you wish to have a hard and ill-natured face? Then while young fill the mind with hard and ill-natured thoughts. Do you prefer a face that shows kindness and honesty? Then cultivate a pleasant disposition, show kindness, and be honest to all. Let the mind be filled with only those thoughts which are true and noble and kind.

General Exercise. Exercise should be taken out of doors as much as possible, since pure air is of the greatest importance. No one need fear taking cold out of doors in the winter season if he keep active while in the cold air and come in the warm house to rest.

Any exercise is too violent which leaves the body "all tired out." It not only makes one unfit to do work of any kind, but also is likely to be injurious to the nervous system. Healthful exercise brings a restful feeling and a desire to work, and insures refreshing sleep.

When and how to exercise. One should not exercise vigorously either just before or just after a meal. We should take some kind of exercise each day. Walking to and from school is not enough, neither will it do to study all the school days, and then play all day Saturday.

The ball, the hoop, and lawn tennis are good for the summer days ; the sled, the skates, and snowballs make good sport for winter. It is true that any work which brings the muscles into play develops and strengthens them ; yet "all work and no play makes Jack a dull boy."

The Muscles must rest. The muscles need rest, and nearly all of them get complete rest when we sleep. But the heart, which is a great hollow muscle, keeps beating away at night as well as during the day. Yet we shall learn that even it has its time of rest, for there is an instant of complete rest between its beats.

General Rule. If exercise be taken for the health, it must be remembered that in order to be most beneficial, the mind must be actively interested in what is being done.

ALCOHOL AND MUSCLE.

Men have devoted a great deal of time to find out just what effect alcohol has on the muscles of the body. They know that when we are hungry and the body is weak, if we take good, nourishing food, we feel strong again. They know, too, that we cannot at once make ourselves strong by eating a great deal.

Let us think of a man who is hard at work in a shop, using his muscles many hours each day. He eats three hearty meals a day, and feels well and strong. Now suppose he wants to work twice as hard one afternoon as he usually does. Do you think he could do it more easily if he should eat two big dinners? Most certainly not.

But when men began to use alcohol, they said: " Here is something that will enable us to work twice as hard as usual, and will not let us get tired."

Is this true? Does alcohol actually make one stronger? Does it enable one to do more work, and not suffer from it in any way? Let us wait a moment before we answer.

Have you ever been very sick? If so, when you were getting better, and were once more walking about, have you not felt as well and strong as ever? Yet when you attempted to lift anything, or to run, you found then that your feelings were no safe guide. You were weak, and could not do what you expected.

Now, our workman in the shop tries a glass of something that contains alcohol, as beer, or whiskey, or brandy, and he says that it makes him feel so much stronger. The real question is this: **Is he any stronger?** Are his feelings a true test?

If alcohol will make a man stronger, and do him no particular harm, then we must all vote it a good thing; but as a result of the most thorough experiments on the lower animals, even on animals as small as the frog, and on the higher animals, even on man himself, it is proved beyond a doubt that both small and large doses of alcohol **reduce the power of the muscles.** We mean by this that if the system be under the influence of alcohol, the muscles will not contract as much as they should.

Our workman, then, was not made stronger by his glass of liquor, he was not kept even as strong. **He was actually made weaker.** He felt as though he could do more work: the truth is, he could not do as much work as he could have done without the liquor. The workman made the mistake of supposing that mental excitement is increased strength.

It is a well-proved fact to-day that alcohol weakens the muscles. The manner of walking, the bent form, the clumsy fingers, and the thick tongue of the man under its influence, all show that his muscles have nearly lost their strength, and that only a little more would be required to make them so weak that they would be unable to do their work, and his body would become completely helpless. Everybody says that a man may take enough strong drink to weaken his muscles and render him unable to walk. The scientific student goes further, and says that even one glass weakens the muscular power.

It is said that some forms of drink, as beer and ale, make the muscles larger. They certainly do tend to make some persons fleshy. But to grow fat is very different from growing strong. To load the body with a great quantity of fat is positively injurious. Beer and ale tend to make an excess of fat; this hinders the proper action of the muscles, and may seriously interfere with the action of many of the organs of the body.

A fatty heart is often the cause of death among beer-drinkers. The fat accumulates to such an extent in the muscle of the heart that it cannot act, and death follows.

As a rule, very fleshy people are neither as strong nor as healthy as those who have much less fat and more hard muscle.

Tea and Coffee. Neither tea nor coffee increases the strength in any way.

Tobacco. Tobacco never gives strength to the muscles, while it may affect some muscles, as the heart, in a very serious way.

SUGGESTIONS TO TEACHERS.

1. A muscle, with its tendon, may be procured at the market. The examination of the leg of a fowl will show how the muscle ends in a fine cord, or tendon. Let the pupil see that the tendon is harder than the muscle; it is made of hard connective tissue, and cannot contract like muscle.

2. The voluntary muscles are represented by the ordinary lean meat of the market. Their action is shown by various movements. Feel the front of the arm while the hand is raised. In a few cases muscle is not attached to bone. To illustrate this, have the letter U pronounced. A voluntary muscle which forms the fleshy part of the lips contracts. Follow this with the letter Y, and the muscle will relax.

3. The heart is the best illustration of an involuntary muscle.

4. Boiled corned beef shows the muscle falling apart into small bundles. These may be still further torn apart. Call attention to the fact that the microscope would show the smallest of these little fibres to be made of still smaller fibres.

5. Interesting experiments may be made to show that expression is caused largely by the contraction of muscles. Contract the muscles of the forehead, as in scowling, to show displeasure. Contract the muscles that draw the corners of the mouth up, and there is an expression of pleasure; while the opposite effect is produced by drawing the corners down.

CHAPTER X.

OUR FOODS.

The Body is wearing out. As the movements of machinery tend to wear it out, so every movement of the body causes some tissue to waste away. Our bodies are constantly wearing out. We must, therefore, make a careful study to find out what we should eat, in order to furnish new material to take the place of that which is worn out; and one of the most important subjects for our consideration is what foods will best keep us in health, and what means we should use to restore us to health when we are sick.

Varieties of Food. Some animals appear to be almost constantly taking small quantities of food. The common canary-bird keeps very busy cracking seeds and swallowing the kernels. There are others that will eat enormous quantities of food at one time, and afterwards lie quiet and perhaps sleep for days. Still others, as the dog, do not chew their food; they tear it just enough to enable them to swallow it, and that is all. The cow chews her food only a little at first. She gathers the grass very fast, and swallows it as quickly as she can. Then when she is resting her body, she brings the food back to her mouth and more thoroughly chews it. Some animals can live on one kind of food. The horse will live on fresh or dried grass, and he never eats meat of any kind. The squirrel likes the kernels of nuts, and does not care for grass. But man makes use of the products of the vegetable and animal kingdoms, and even of the mineral kingdom.

Tastes differ. If we travel through various countries we find that the people of one land eat substances which are declared by the people of another to be very repulsive. There are persons in the world who eat moths and bees, rats and mice. The savage man, as the Indian, eats nearly all his food in its natural state; he grinds the grain into a coarse powder, mixes a little water with it, and cooks it over the fire. The civilized man, however, first takes the grain to the mill, and there not only gets it ground very fine, but also has certain objectionable matters removed.

FOODS FROM THE MINERAL KINGDOM.

The two principal articles we use from this kingdom are water and salt.

Salt. Salt is found in every tissue in the body except the enamel of the teeth. As it is so generally distributed throughout the body, we must take it with our food, or suffer greatly. Salt is naturally present in nearly all the foods we use, but only in quantities so small that there is not enough furnished in this way to meet the demands of the system; therefore we add it to our food.

Its Uses. Salt supplies a demand of the body, and gives flavor to the food, thus making it more pleasant to us; it also causes the digestive juices to flow more freely, and stimulates our appetite.

It is Necessary. Experiments have been made on animals to ascertain the effect of depriving them of salt. It was found that their hides became rough, their eyes grew dull, they were less active, and at last they lost their health and strength.

A Natural Demand for it. The farmer knows how quickly his sheep will come at his call if they have learned that by so doing they can get some of this necessary food. Cattle will eat the coarsest kind of fodder if only they taste the salt that has been sprinkled over it. This shows the natural demand of the system for this mineral substance. In some parts of the world salt is very valuable. Fifty years ago it was expensive even in our own country. Sometimes it would take a whole load of wheat to purchase a single barrel of it. At the present time it is very cheap and plentiful. In western New York a dollar will buy all the coarse salt that a horse can draw away.

Water. Nearly three fourths of the weight of the body is composed of water. If a person weighs 120 pounds, it is estimated that 85 pounds will be water. This fact alone is sufficient to show how important it is that we should have the purest water to use.

A Natural Demand for it. The craving for water is greater than for food, and those who have been deprived of it describe their sufferings as terrible. A person will die sooner if deprived of water than if deprived of food.

Large Quantities taken. We little suspect how much water we take into the body each day. Every kind of food we use contains it. When we eat beef we take over one half its weight in water. Potatoes, that look so dry and mealy, consist of three fourths water; even dry sugar contains a large amount of it, while milk consists of nearly nine tenths water. So if we do not purposely drink water, we still use a great deal of it. A healthy man takes with his food and as drink on an average about two quarts of water each day, and it is necessary that it be taken in these large quantities; for without it the blood and the secretions could not properly perform their work, and even the muscles and tendons would be made to suffer.

Other Substances with Water. The water we use for drinking always contains some mineral matter and gases, and occasionally some vegetable matter. When the mineral matter, as lime, is not in excess, it is useful to the system. As lime is so important in the formation of the teeth and bone, water that has a small amount of lime in it must be regarded as beneficial, especially when the tissues are developing during early life. When water is carried through lead pipes it may dissolve enough of the lead to act as a poison when taken into the body. We should consequently never drink water that has stood in lead pipes. If such pipes are used, the water should be allowed to run through them constantly. In this way the danger is greatly reduced.

A Cause of Sickness. Impure drinking-water is the cause of an immense amount of sickness, and thousands of deaths occur each year as a result of its use. The fact that water looks clear and has no odor gives no reason for supposing that it may not have something dissolved in it that will produce some terrible disease. Filth is the cause of many diseases, and of typhoid fever in particular. If we wish to escape this disease, we must be sure that our drinking water is free from filth of every kind.

Where is the Well? There is a very general idea that it is only necessary to dig a hole anywhere in the ground in order to get good drinking water. This is a great mistake. The well should never be in the house, it should not be near the barn, nor should it be near a place where there is any refuse matter. We must not forget that the germs of disease may be carried through the soil for a considerable distance, and in

this manner reach a well that is many feet away, especially if the soil is sandy, or slopes towards that well.
A well should be at least fifty feet from any place
whence filth might get into it.

To purify Water. In cities, where good water is not
easily obtained, it is a wise plan to boil the water
before it is used for drinking. After boiling, it can
be set aside to cool in the winter, and placed in the
refrigerator in the summer. Ice should not be put into it, but around it.

Too much Water injurious. It is not wise to drink too
much water with our meals, as it dilutes the digestive
juices, and is a frequent cause of stomach troubles. It
is certainly a very bad practice to chill the stomach by
putting ice-cold water into it.

FOODS FROM THE ANIMAL KINGDOM.

At the head of the list of these foods, for ease of
digestion and for concentrated nourishment, we must
place eggs. When soft boiled in the shell, or when
dropped into boiling water, they are acceptable to the
most delicate stomach, and contain the most nourishing
materials. Oysters also are very nourishing, and they
are easily digested, especially when eaten raw.

Beef is undoubtedly the best meat for general use.
Tender beef, properly cooked, is easily digested, and
agrees with most persons. Mutton is nearly as good as

beef; but there are many persons with whom it does not agree, or to whom its flavor is not agreeable. Veal is neither as easily digested nor as nourishing as either mutton or beef. Lamb is more easily digested than veal, but not so nourishing as mutton. Pork is used by a great many persons, and to those who have strong digestive powers it appears to do no harm. It is, however, very difficult to digest, and should never be eaten by those whose stomachs are weak. Lobsters and crabs are exceedingly difficult to digest, and should be avoided by invalids.

FOODS FROM THE VEGETABLE KINGDOM.

The principal grains used as foods are wheat, corn, oats, and rice, all of which contain a large quantity of starch. These foods are very important to mankind, millions of human beings eating scarcely anything else.

While starch is the principal substance in these grains, yet mineral matter, oil, and fat are present also. Wheat stands at the head as the most useful of the grains. Oatmeal contains starch and a good supply of mineral matter; it is a wholesome food, easily digested, and to most persons agreeable. The potato is the most generally used vegetable; it is composed almost entirely of starch and water.

Peas and beans are very nourishing, and are so valuable as foods that they would be used much more than they are, were it not for the fact that they consist of so solid matter that they are not easily digested.

Turnips, cabbages, parsnips, onions, etc., are added to our list of foods, that we may have a suitable variety; but they are not very nutritious, nor are they easily digested.

Apples, peaches, and other fruits are useful to us in many ways. The acids they contain stimulate the appetite and excite the flow of the gastric juice, while the water they contain serves to quench the thirst; but they are not very nourishing. If uncooked, they are most wholesome eaten before meals and in the early part of the day; but if cooked they can be eaten with any meal. Dried fruits and nuts are not easily digested, and should be eaten sparingly.

OTHER FOODS.

Sugar. Sugar forms an important article of our diet, and a proper amount of it should be used. It makes certain articles of food more pleasant to the taste, but when eaten in too large quantities it is likely to cause trouble with the stomach, and thus impair the health. If we desire to satisfy our natural appetite for sweet things, it is better to use home-made candies than the impure and highly colored candies from the stores.

Milk. Milk must be regarded as a perfect food; no ideal food could surpass it. It contains in a digestible form all the elements most necessary for the support of the body. Many forms of stomach disease are cured by a diet of milk alone, and it is given by physicians in fevers and other diseases. Great care should be taken to keep milk sweet and pure, as it will readily absorb gases, — a fact easily proved by placing a bunch of onions near a dish of milk in a closed box. The milk will soon be tainted by the gas from the onions, and will show it in taste and odor. Milk should be kept in a clean room, where the air is always pure and sweet.

If the value of milk as a food were better understood it would be much more largely used. For adults it may be used as a drink with the ordinary meals. During the cold weather of winter it may be taken warm with the breakfast, and cool with the other meals. Even during the heat of summer it should not be taken into the stomach ice cold. Cool milk may be used, but large quantities of iced milk are certainly injurious, especially when taken with meals.

Cream and Butter. Butter is a most important article of diet. It is valuable because it supplies the body with needed fatty material, and also because it gives flavor to other foods, thus making them the more readily eaten and digested. It is composed principally of the fat of milk. Under the microscope milk is found to consist of a large number of minute oil-drops floating in water. We know that oil is lighter than water, so that when the two are shaken together and then allowed to stand, the oil will rise to the top. So when milk is allowed to stand, the oil will rise to the surface. This oil is called cream. Churning cream is simply beating these minute oil-drops, or the fat of milk, into one solid mass.

Buttermilk, Skim milk, Cheese. Buttermilk is a wholesome, cooling drink in the hot summer time. Skim milk contains a small amount of fat and some of the mineral matter found in milk. It is more nutritious than buttermilk. Cheese is used principally as a side dish to promote the appetite or to give variety to our table supply. It is classed with the foods most difficult of digestion.

QUESTIONS.

1. What is said about the wearing out of our bodies?
2. Tell about the food and manner of eating, of some animals?
3. Do all peoples like the same kinds of food?
4. What can you say about salt in the tissues of the body?
5. What are the uses of salt?
6. Is salt necessary?
7. How do we know that sheep and cattle are fond of it?
8. What proportion of the weight of the body consists of water?
9. What can you say about the demand for water
10. What is said about the large quantity taken?
11. What substances are usually found in drinking water?
12. What is said about lime in water? About lead?
13. Does impure drinking water ever cause sickness?
14. What caution should be used in choosing a place for a well?
15. How can water be easily purified?
16. Name some foods from the animal kingdom.
17. Tell what is said about eggs, oysters, and beef.
18. Name some grains and fruits used as foods, and give their value.
19. What is regarded as a perfect food? Why?

CHAPTER XI.

COOKING.

Raw Meats. Raw meat is used by a large number of people. A fondness for it is said to be readily acquired. Raw meats are not so easily digested as cooked meats. They are generally eaten after being smoked or dried.

Cooking is necessary. It is said that it is impossible to find a race of men so uncivilized that they do not cook a part of their food. We are educated to believe that it is absolutely necessary to cook some kinds of food. Cooking is necessary because it brings out flavors that are agreeable, and thus increases the appetite and stimulates digestion. It either softens the article, or aids in dividing it into small particles, and thus promotes digestion. Improper cooking, however, will make the purest and best articles of food harmful to us.

Broiling. A tender piece of beefsteak, carefully broiled, contains great nutritive properties and is easily digested. Broiling is the best way to cook meats; next to this roasting, and then boiling.

Boiling. In cooking meats we should remember that the natural juices should be retained in them as much as is possible. This can be done by making the meat very hot at once, in order that it may be hardened on the outside, thus forming a crust through which the juices cannot escape. Therefore all meats that are to be boiled should be put in boiling water, and roasts should be placed in the hot oven.

Frying. Frying makes meat hard, and difficult to digest. If frying is to be done at all, the fat should be boiling hot before the food is put into it. Then an outer crust is formed at once, and the oil does not pass into the meat.

Making Soups. When it is desired to make soups or beef-tea, for the purpose of using the juices and not the meat, the meat should be cut in pieces and placed in cold water, and the water allowed gradually to come to a high temperature. Mutton broth prepared in this way is very nutritious and easily digested.

Eggs. Eggs are cooked in many ways. For the most delicate stomach there can be nothing better than a fresh egg broken into boiling water and cooked just enough to make the albumen, or white, solid. A soft-boiled egg is also very nourishing and easily digested. Hard-boiled eggs should be eaten only by those who have strong powers of digestion.

Vegetables. Vegetables should be thoroughly cooked. As a rule, they are not cooked enough. The practice of frying them does not give a wholesome food. Some vegetables, as lettuce and radishes, are eaten without cooking. If taken in moderate quantity, they serve as a relish, and stimulate the appetite.

The Starchy Foods. The various starchy foods, as corn-starch, rice, and oatmeal, should be boiled a long time, that they may be properly acted upon by the digestive juices.

New Bread. Newly baked bread is difficult to digest, because it is likely to form a soft, pasty mass in the mouth, so that when it reaches the stomach it is a solid lump into which the digestive juices cannot easily enter. Light, sweet bread is always acceptable to the taste, and is highly nutritious.

Pies and Cakes. Pies and rich cakes are not wholesome foods for delicate stomachs, and should be used only in moderate quantities by any one. The pie crust contains too much fat, as lard or butter, and the cakes have too much sugar and butter in them to make proper foods. They should never be taken in large quantities.

Cooking is an Art. Cooking is a great art, and full of hidden secrets. It is founded upon laws of chemistry and physiology, and a knowledge of these sciences is necessary to understand fully its mysteries.

What Food shall we eat? No rule can be given either for the kind or the quantity of food we should eat. We must learn what are wholesome foods, and how they can be spoiled by improper cooking; then each one must decide for himself what he will eat. It can be given as a rule that a general mixed diet is the best.

A Cause of Sickness. Doubtless one of the great causes of sickness is due to eating too much and eating things that are harmful. Yet even this will depend upon the habits of life. An Esquimau can eat from twelve to fifteen pounds of meat a day, besides a large quantity of fat. An Arctic explorer says he saw an Esquimau eat thirty-five pounds of meat and several tallow candles in the course of a single day. Some persons require but a small amount of food to keep them alive and in good health, and others have lived a number of days without food.

Plenty of Food for Children. Children should have plenty of good food. They take much exercise, and their bodies are growing rapidly. For these reasons Nature gives them a strong appetite, and she expects us to satisfy it with a good supply of wholesome food.

IS ALCOHOL A FOOD?

Is alcohol a food? Does it nourish the body in any way? Does it build up the body? Does it satisfy our hunger? Does it quench our thirst?

If alcohol is an important food; if it strengthens the laboring man; if it aids the student in his work; if it answers a demand of the system, — then we shall have to declare it is a good thing.

But we know that it will not do this. There is no water in it to supply this much-needed fluid to our bodies. On the contrary, it has the power to take water from other things; therefore when in the body it is capable of taking water from parts where it is most needed. It has no mineral product, as lime or salt, that might be of use to us. It has nothing in it, like the meats of the animal kingdom, which can be readily changed into the flesh of our bodies. There is nothing in it like the products of the vegetable kingdom, as starch or sugar, which are so highly prized as valuable parts of our food. We can find nothing about it that gives us any idea that it is a food.

" But," it is said, " we have to eat to keep the body warm; and alcohol raises the temperature of the body, and thus acts as a food." This is a great mistake.

We are asked: " Is it not true that a glass of whiskey or brandy will enable us to endure the cold weather of winter much better ? " This we shall answer more completely when we come to study the temperature of the body; but in the mean time we shall quote from one of the most prominent medical writers of the day: "In nearly all the cases of death caused by exposure to cold that I have known or heard of, it was found that the persons so dying had taken some alcoholic drinks, not always in large quantities, before going out into the cold. So well is this known by people in the northwest of America and in Canada that they will seldom take a glass of spirits when they are likely to be exposed to extreme cold."

We must remember that alcohol is the active principle in all spirituous drinks. Men drink beer and ale and wine for the effects of the alcohol they contain. There is one glass of alcohol in about every twenty glasses of beer. Alcohol is not in any sense a food.

Is there no nourishment in beer? One of the greatest of our authorities on foods says: "There is more nourishment in the flour that can be put on the point of a table knife than in eight quarts of the best beer."

"Then," it is asked, "as beer increases the fat of the body, is it not, therefore, a valuable food?"

The answer to this question explains the terrible effects of alcohol on many of the organs and tissues of the body, and gives the cause of a number of diseases. The natural conclusion would be that if beer makes one fleshy, it must be a valuable food. But a more careful study will show this conclusion to be a wrong one. At the same time that the tissues beneath the skin are becoming changed into fat, pushing the skin out and making all the wrinkles disappear and the face look plump and round, just the same thing is going on in the deeper and more important tissues. The strong, hard muscles are changing into fat; they are growing weak and soft and flabby. The liver is changing into a mass of fat, and even the heart is becoming fatty and weak. The more important tissues and organs in the body lose their proper structure and become more or less changed into fat. Of course this often causes the most serious trouble, and not infrequently is the cause of death itself.

The fleshy body caused by the continued use of large quantities of beer is a diseased body. Strong muscles, a healthy heart, and a sound liver are not found in such a body.

QUESTIONS.

1. Are raw meats easily digested? How are they generally prepared for eating?
2. Why is cooking necessary?
3. What method of cooking beefsteak is the best?
4. How should we boil and roast meats? Why so?
5. Is frying a good method of cooking meat?
6. How are soups best made? Why so?
7. Give the best ways of cooking eggs.
8. What about the cooking of vegetables?
9. Tell about the cooking of the starchy foods.
10. Why is new bread difficult to digest?
11. Are pies and cakes especially good foods?
12. Is sickness ever caused by overeating?
13. Upon what may overeating sometimes depend?
14. Do children require much food?
15. Is there any water in alcohol?
16. Any mineral matter in it?
17. Anything about it like a food?
18. Is there any nourishment in beer?
19. What can you say of the body which is made fat by beer?
20. Are healthy organs found in such a body?

CHAPTER XII.

DIGESTION.

WE eat and drink because we are hungry and thirsty; and if our bodies are in a healthy condition, hunger and thirst may be taken as safe guides.

Hunger. The sensation of hunger is generally said to be in the stomach; but it is not confined to any place: it is a call of the whole system for food.

Thirst. Our throats seem to tell us when we are thirsty; but this is a call of the whole system for liquids, and is not confined to any one part.

Digestion. Digestion is the process of preparing the food that has been taken into the stomach in order that it may be absorbed. The food we take must be dissolved and changed before it can be absorbed, or taken up, by the vessels and carried by them to all parts of the body.

Why is Food necessary? Our bodies are constantly wearing out, and for this reason alone we must perish if we do not furnish a fresh supply of food. Then, too, in growing persons we ought not only to furnish new ma-

terial to take the place of the old, but also enough more
to make new tissues, so that the body may properly
increase in size.

What becomes of the Old Material? A healthy body
must not only furnish new material, but also get rid
of the old, worn-out material. If we neglect to supply food, we shall soon starve ; while if our bodies
fail to remove the old, worn-out material, death will
speedily follow : for this old material acts as a poison
to the whole system. The skin, lungs, kidneys, and
other organs all aid in carrying off this poisonous
substance.

The Alimentary Canal. Digestion is carried on in the
alimentary canal. This is about thirty feet in length in
the adult. There is an enlargement in its upper part,
called the stomach. Near the alimentary canal there
are some glands, called the liver and the pancreas.
Small canals, or ducts, lead from these to the inside of
the canal. The secretions, or juices, from these glands
are poured into the canal, in order that they may there
mix with the food and make certain changes in it so
that it can be absorbed.

Mastication. The first act of digestion is called mastication, or chewing. As soon as Nature thinks it is time
for us to take solid food, she furnishes us with teeth to
chew it. A child has but twenty teeth, — ten in each jaw.
The first teeth appear about the sixth or seventh month,

and the last about the end of the second year. These teeth do not stay long, for the second set begin to appear about the fifth or sixth year. These come one by one until the child is twelve or thirteen years of age, when all will be present, except the wisdom teeth. These usually do not appear until the person is twenty or twenty-five years of age. In Fig. 20 the position of some of the second teeth is shown. They are formed in the jaw just beneath the others, and gradually rise up to take their places.

FIG. 20.

Importance of the Teeth. The teeth are very necessary to digestion. They break up the food into fine particles and mix it with the saliva. They are very important in another way : they add very much to the personal appearance of any one, especially when they are regular and white. A set of filthy and decayed teeth is very repulsive.

Care of the Teeth. We must not neglect our teeth in any way if we wish to preserve them and escape the pains of toothache. The teeth should not be picked with a pin or with any hard substance. Nuts should never be cracked between them; they should be thoroughly cleaned at least once each day. It would be better to clean them after each meal and at bedtime. Use a small, soft brush at least once each day, and as soon as a cavity appears, consult a dentist and have it filled.

The Teeth vary in Shape. Our teeth are of various shapes, because they have different kinds of work to perform. The front ones are sharp, for cutting, hence called incisors; while the back ones are large and uneven for grinding, and these are called molars.

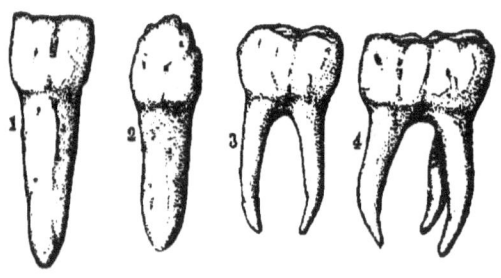

Fig. 21. The teeth of an adult. (1) an incisor, or cutting tooth; (2) a canine, or eyetooth; (3) a molar, or grinding tooth, of the lower jaw; (4) a molar of the upper jaw.

The Inside of a Tooth. By breaking open any tooth a cavity will be found within it. (Fig. 22.) In the living body this cavity is filled with nerves and blood vessels. Fig. 23 shows that each tooth has its own blood vessel and nerve. It is when these are diseased that a tooth gives us intense suffering.

The Saliva. There is a constant flow of a liquid into the mouth. It is called the saliva. While the jaws move and the teeth are doing their work, the flow is greatly increased. The saliva comes from some glands near the tongue, and also from two large glands, one just in front of each ear. In health we are not conscious of their whereabouts; but when they become inflamed we know all about it. The mumps are an in-flammation of the large glands.

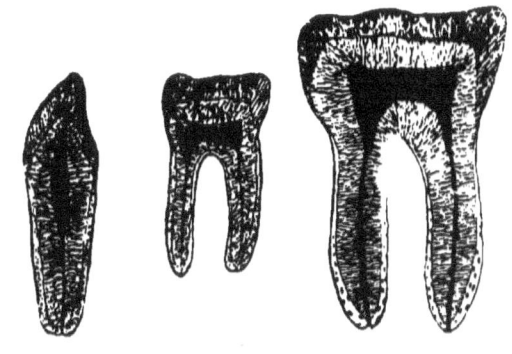

FIG. 22. Three teeth broken open to show the cavity on the inside. Surrounding the top of each tooth is the enamel. The two teeth to the left are but little larger than natural; the one to the right is more highly magnified.

Slow and Thorough Mastication. One of the most frequent causes of trouble with the stomach is too rapid eating. The solid foods should be chewed very finely, and all food well mixed with the saliva.

The Object of Mastication. The saliva has the power of changing the starch of our food into sugar. As all starch must be thus changed before it can be taken up by the blood vessels, it follows that the action of this fluid is important. However, not all the starch is con-verted into sugar while it is in the mouth. The food is in the mouth such a short time that only a part of it is affected. The rest is changed at a later period farther

6

down the alimentary canal. The principal object of mastication is to finely divide the solid foods and to moisten them for swallowing.

Too much Fluid is injurious. It is a harmful practice to wash down the food with large quantities of water or tea or liquid of any kind. The saliva will furnish moisture enough as a rule; for there are from one to three pints of it poured into the mouth of an adult

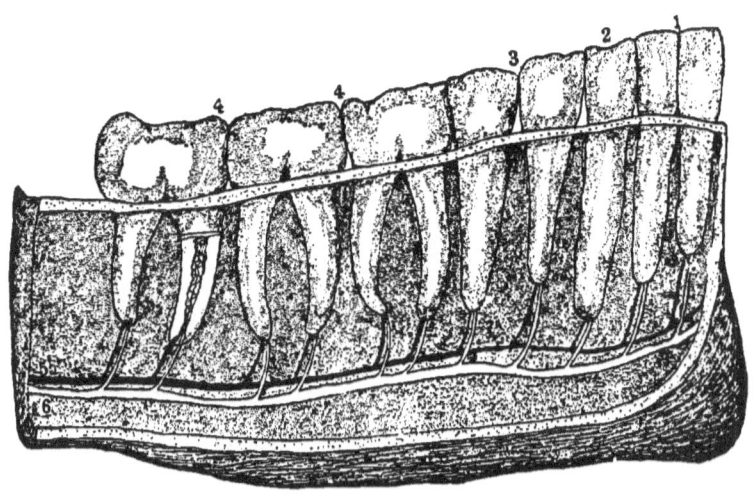

FIG. 23. A side view of the lower jaw with the outer walls of bone removed, showing the teeth in proper place. (1) the two incisors ; (2) the canine ; (3) the two bicuspids ; (4) the three lower molars (the last molar is sometimes called the wisdom tooth) ; (5) a blood vessel ; (6) a nerve.

each day. There is no harm in using a moderate amount of drink with our meals, if it be of the right temperature. Iced drinks should not be used under any circumstances, neither are very hot drinks much better. A glass of milk or a single glass of water should be sufficient.

Swallowing. By the act of swallowing, the soft mass is carried from the mouth down the œsophagus, or gullet, into the stomach. Here very important changes take place.

QUESTIONS.

1. Is hunger confined to any particular place in the body?
2. What is thirst?
3. What is digestion?
4. How must the food be changed before it can be absorbed?
5. Why is food necessary?
6. Why get rid of the old, worn-out material?
7. What parts aid in carrying it off?
8. Describe the alimentary canal.
9. What glands are near this canal?
10. What is the first act of digestion?
11. Give some facts about the teeth.
12. Of what general use are the teeth?
13. Tell how we should care for the teeth.
14. Describe the different shapes and uses of the teeth.
15. What is found inside of a tooth?
16. What is the saliva ; and where does it come from?
17. What is said about thorough mastication?
18. What is the effect of saliva on starch?
19. What is the principal object of mastication?
20. What is said about drinking too much with the meals?

CHAPTER XIII.

DIGESTION IN THE STOMACH.

The Stomach. The stomach of an adult is nearly a foot in length, and three or four inches in diameter. It has a firm outer wall of involuntary muscle, while the inside consists of a delicate membrane, called the mucous membrane.

Fig. 24 shows this membrane arranged in folds, or wrinkles. When the stomach is well filled, these folds spread out and disappear.

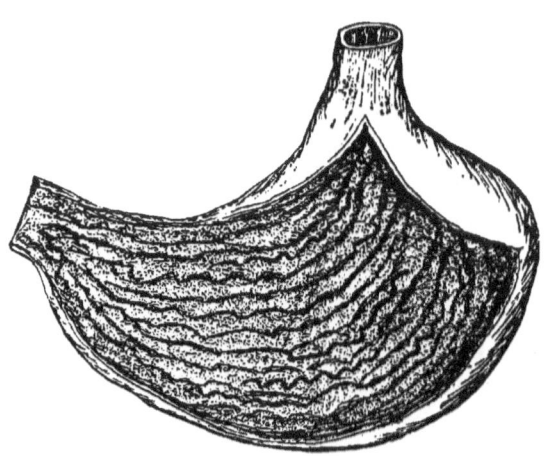

Glands of the Stomach. This membrane has glands in it so very minute that they cannot be seen without a microscope.

Fic. 24. View of the inside of the stomach. The front walls have been cut away.

Fig. 25 shows how a vertical section of this membrane looks when examined with a small magnifying glass. The inside of the stomach is at

the top of the figure, and
at the bottom is the wall
of muscle. Nine of these
little glands are seen all
opening at the top, or on
the inside of the stom-
ach. Fig. 26 is the sec-
ond gland from the right,
more highly magnified.
Fig. 27 is the third gland

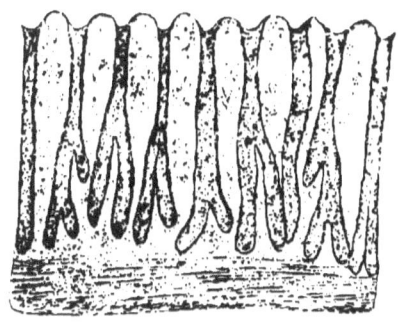

FIG. 25. A cross section of a small portion of the walls of the stomach, slightly
magnified, showing the glands.

from the left, as seen under a more powerful microscope.

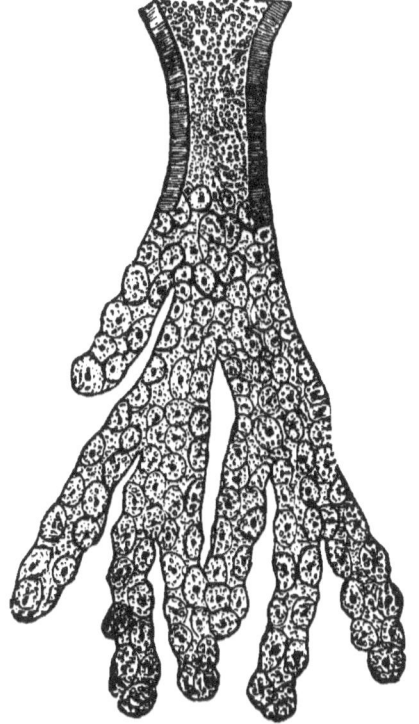

FIG. 26. One of the glands of the stomach, as seen with a microscope.

FIG. 27. A gland of the
stomach, highly magnified.

The Gastric Juice. The round bodies, or cells, seen in these glands make a juice, called the gastric juice. Just as soon as food reaches the stomach these cells begin to pour out this juice, which is to change much of the food. At the same time the muscles on the outside of the stomach begin to contract, mixing the food thoroughly with the juice.

The gastric juice changes certain of our foods so that they can be taken up by minute vessels and carried to all parts of the body. Not all the foods we take are digested in the stomach by this juice ; some of them pass out of the stomach as they entered it, and are digested in the canal near it. The oily or fatty foods, and all the starchy foods that are not changed in the mouth, are digested after they have left the stomach ; but such foods as the lean meats and eggs (albuminous foods) are digested in the stomach. When the action of the gastric juice is completed there remains a grayish fluid, called chyme.

Pepsin. The fact that solid meats and some other foods can be digested in the stomach is largely due to the presence of a substance in the gastric juice known as pepsin.

The Time required for Digestion. A few hours after food has been taken, the stomach is again empty. The time required for the stomach to complete its work depends upon the kind and the amount of food, the liquids that are taken during the meal, the health of the person, and other conditions.

The following table shows that digestion is completed in from one to five hours : —

Easy of Digestion.	h. m.	More Difficult.	h. m.
Rice, boiled	1.00	Potatoes, boiled . . .	3.30
Apples, sweet, raw .	1.30	Oysters, fried	3.30
Milk	2.00	Eggs, hard boiled . . .	3.30
Cabbage, raw . .	2.00	Pork, broiled	3.30
Oysters, raw . .	2.30	Beef, fried	4.00
Potatoes, baked	2.30	Cheese	4.00
Chicken, boiled	2.45	Cabbage, boiled . .	4.30
Eggs, soft boiled	3.00	Duck, wild, roasted . .	4.30
Custard, baked	3.00	Pork, fried	4.30
Beef, broiled .	3.00	Pork, roasted	5.15

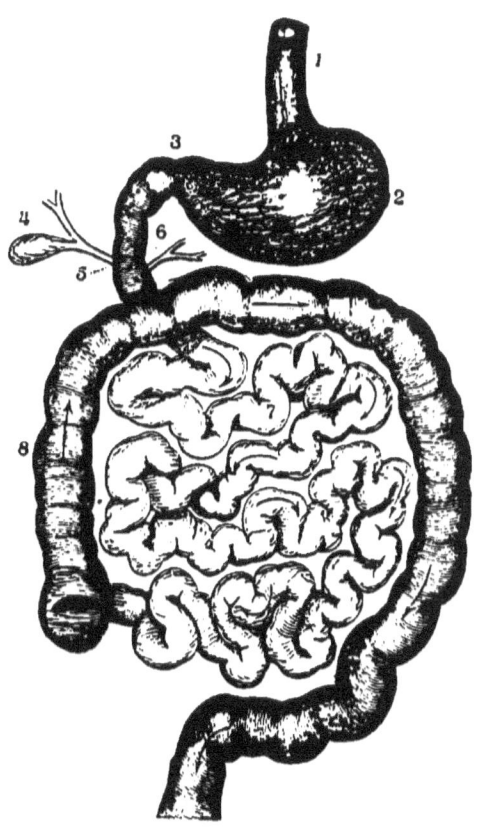

The time required for foods to digest was ascertained by a Dr. Beaumont, who experimented upon one of his patients. This patient, Alexis St. Martin, had an opening from the surface of the body directly into the stomach, as a result of a gun-shot wound. Food could be introduced through this opening into the stomach, and the whole process of digestion could thus be carefully studied.

Fig. 28. (1) The œsophagus ; (2) the stomach ; (3) the pylorus ; (4) the gall bladder ; (5) the duct carrying bile to the intestine ; (6) the duct from the pancreas ; (7) the small intestine : (8) the large intestine.

The Pylorus. The name of the end of the stomach through which the food will pass out is the pylorus. The food will not pass through this until the gastric juice has had time to act upon it. Then the pylorus will open, and by the contraction of the walls of the stomach the food is pushed through it into the upper part of the intestines.

Fig. 28 shows the general arrangement of the alimentary canal. Just below the stomach are the small canals, or ducts, coming from the pancreas and the liver. A portion of the intestine where these ducts enter it is shown in Fig. 30.

HYGIENE OF THE STOMACH.

Liquids at Meal-time. If we should hastily drink a glass of iced water during a meal, the cold might be sufficient to stop the action of the glands of the stomach, and thus check the forming of the gastric juice. The glands would not begin their work again until they had recovered from the shock of the cold. Thus digestion would be prolonged, and some form of stomach trouble would be likely to follow. Too much liquid of any kind is harmful, because it dilutes the gastric juice, and therefore weakens it.

Eat slowly. The gastric juice will not dissolve, or digest, the solid foods in proper time if the pieces are too large when swallowed. We should eat slowly, not only because it is good manners, but also because it gives time for the gastric juice to be formed, and to be thoroughly mixed with the food as it is swallowed.

CHAPTER XIV.

DIGESTION IN THE INTESTINE.

THE process of digestion is not completed in the stomach. After the food, or chyme, has passed into the alimentary canal it undergoes a final change.

Within a few inches of the stomach there are poured into the alimentary canal the juices secreted by the liver and the pancreas. These juices change the chyme into a milky fluid called chyle. Fig. 28 indicates the situation of the ducts that convey these juices, and Fig. 30 shows more clearly how they open into the canal.

The Liver. The liver is a large organ principally in the right side of the body, although Fig. 29 shows that a portion of it extends to the left side. In the adult this organ weighs between three and four pounds.

The Bile. The juice made by the liver is called bile. The liver is constantly secreting bile, although there are times when an increased quantity of it is poured into the intestines. The bile may pass directly from the liver into the intestines. or it may first pass into the gall-bladder, and from there into the intestines. The gall-bladder acts as a reservoir for the bile. Fig. 30 shows that the duct from the liver unites with the duct that comes from the pancreas, and that both terminate as one duct, at the point 10.

The bile affects the lining membrane of the intestine

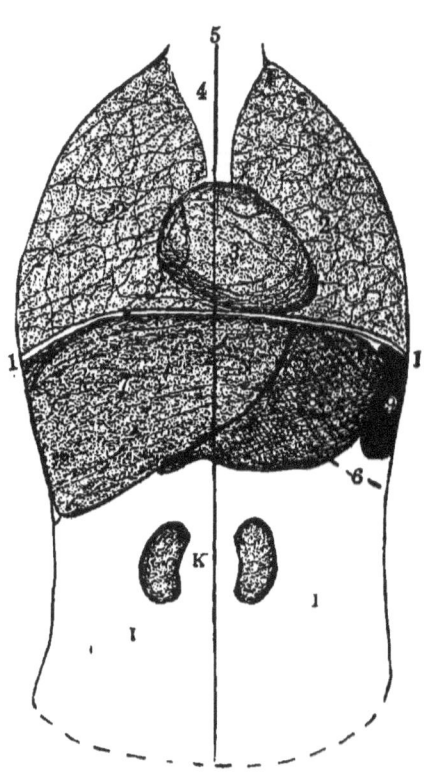

in such a way that it aids in the absorption of the fats. About two and a half pints of bile are secreted each day. If the liver fails altogether to secrete this substance, sickness and death follow. If enough is not secreted, then the whole body is affected, and sickness follows, which in time may alone be sufficient to cause death. When the passage from the liver to the intestine is stopped up in any way, then the bile is taken up by the blood vessels and carried to all parts of the body, making the skin yellow. The person thus affected becomes very ill, and we say that he is jaundiced.

FIG. 29. A diagram illustrating the position of the principal organs of the trunk: (1) is the diaphragm, which divides the trunk into two large cavities, -- the thoracic cavity, above, and the abdominal cavity, below ; (2) the lungs ; (3) the heart; (4) position of the trachea; (5) the median line of the body ; (6) a dotted line, showing the lower border of the ribs; (7) the liver ; (8) the stomach ; (9) the spleen; [consult fig. 30 for the situation of the pancreas ;] (K) the kidneys ; (I) the position of the intestines [consult fig. 28].

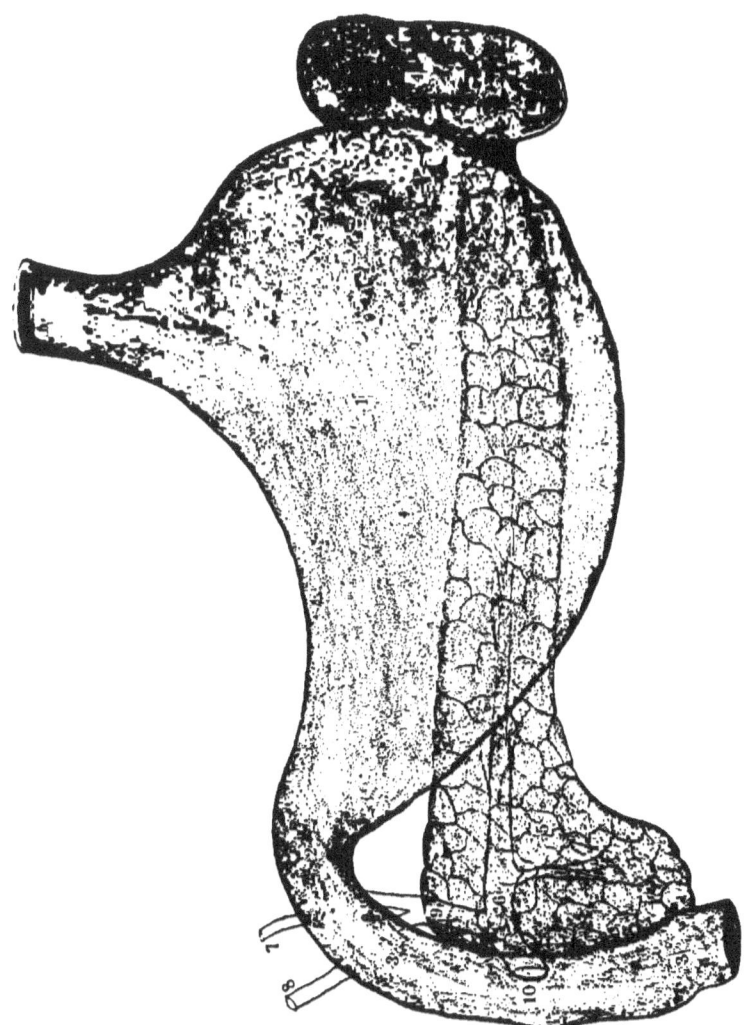

FIG. 30. A diagram illustrating the position of the pancreas and its relation to surrounding parts: (1) the stomach; (2) the pylorus; (3) the small intestine; (4) the spleen ; (5) the pancreas; (6) the duct of the pancreas; (7) the bile duct from the liver; (8) the bile duct from the gall bladder ; (9) the common bile duct, formed by a union of the two bile ducts. The common bile duct unites with the pancreatic duct, and the one duct thus formed opens into the small intestine at 10.

The Liver Sugar. We remember that all starchy foods are changed into sugar before they are absorbed ; therefore all the starchy foods and all the sugar we eat are absorbed into the body as so much sugar. This sugar is carried directly to the liver. The liver makes certain changes in it, and then stores it up in its cells until some time between meals, when it will give it out again to the body. Thus the liver acts as a great storehouse. It takes some of the digested food when there is plenty of it, and stores it up until such time as the body needs it, and then gives it out as so much sugar. This is called liver sugar.

The Pancreas. Just below and under the stomach is the pancreas. This is a slender organ, about six inches in length, which secretes a juice called the pancreatic juice. A duct carries this juice to the intestines. Fig. 30 shows the general shape of this organ, and how its duct unites with the one from the liver. It illustrates how the stomach covers a large part of the pancreas. The pancreas is drawn as though it could be seen through the stomach, in order that its position may be the better understood. To the right of the figure, and at the small end of the pancreas, is the spleen. The small end of the pancreas and the spleen are on the left side of the body.

The pancreatic juice is very important in digestion, as it changes the starchy foods into sugar, and thus completes the work begun by the saliva. This juice alone will digest meats and all foods that the gastric juice can digest. But it does one thing that none of the other juices do, — it digests the fats, and makes them so that they can be easily absorbed.

Glands of the Intestine. There are glands in the walls of the small intestine that secrete a juice which aids in the digestion of foods, especially the starches and fats.

How many Fluids? We have learned that there are five fluids used in digesting the food. First, the saliva; second, the gastric juice; third, the bile; fourth, the pancreatic juice; and fifth, the intestinal juice.

The Quantity secreted. It is estimated that as many as twenty pounds of these juices are secreted every twenty-four hours.

TO HAVE A GOOD DIGESTION.

Eat slowly.
Eat at regular hours.
Eat fruit before meals.
Chew the food thoroughly.
Be cheerful at the table.
Do not eat between meals.
Never eat just before going to bed.
Eat only a few kinds of food at a meal.
Never eat after the appetite is satisfied.
Do not wash down the food with too much drink.

Avoid iced water and iced drinks at mealtime.

Avoid too severe exercise after a hearty meal.

Avoid too much sugar or sweet food; it is likely to ferment in the stomach.

QUESTIONS FOR CHAPTER XIII.

1. Describe the stomach.
2. What does the microscope show in the mucous membrane?
3. What do these glands make?
4. Does the gastric juice change all our foods?
5. What foods does it not change, if any?
6. What is chyme?
7. What is the active substance in the gastric juice?
8. What is the pylorus?

QUESTIONS FOR CHAPTER XIV.

1. What other juices besides the gastric juice?
2. Describe the liver.
3. What is the bile?
4. Into what does the bile pass?
5. What is the purpose of the gall bladder?
6. Of what use is the bile?
7. Is the secretion of the bile important to health?
8. How must the starchy foods be changed before they can be absorbed?
9. Where is the sugar carried?
10. What does the liver do with it?
11. Describe the pancreas.
12. What effect has the pancreatic juice on the foods?

CHAPTER XV.

ALCOHOL, TOBACCO, OPIUM, AND THE DIGESTIVE ORGANS.

Effect on the Stomach of these Drugs. We do not intend to show the effects of pure alcohol on the stomach. Pure alcohol is not ordinarily used ; it would have the most serious effects at once, if taken undiluted. Alcohol is taken in wine, whiskey, beer, or other liquors, and it is the effects of these that we purpose to discuss. The strongest liquors, such as brandy, consist of one half alcohol, while the weaker liquors, such as beer, contain about two tablespoonfuls to a large tumblerful. When an alcoholic drink was given to Alexis St. Martin, it was noticed that the mucous membrane of his stomach suddenly became very red, — as we might expect the eye to become red if pepper were thrown into it. After using alcoholic beverages freely for some days, his stomach looked very red and inflamed, and the gastric juice was thick and unnatural in appearance. The very first effect of these beverages on the mucous membrane of the stomach is to cause an increased flow of blood to it. Now, a most important result follows this. The glands of the stomach secrete an unusual amount of gastric juice. This is so important an effect that if we could only stop our investigation here, we should be justified in believing these liquors are of great benefit, causing a more abundant flow of the gastric juice. But other effects follow. What are they?

Let us quote from one of the greatest living authorities on this subject. In a work published on the action of various drugs on the human body, written for medical men alone, this writer says: " The first effect is an increased flow of blood, the next is an increased flow of gastric juice. After this stimulation of the glands for a few times, the most serious changes occur in the mucous lining of the stomach, and in the stomach glands as well. The membrane becomes constantly red, or inflamed, and therefore secretes an imperfect juice; later the stomach glands become smaller, and there is great difficulty in digesting the food." All this gives rise to a disease known as gastric catarrh. This is a slow inflammation of the stomach. The inflammation makes heat, and this heat gives rise to thirst and to a peculiar faint or sickening feeling, to quench which more liquor is used. For a short time this deadens the feeling, and the man thinks his drink has helped him. He argues that he has a bad stomach, and he drinks only a glass or two of beer or spirits as a medicine; for he notices that after each drink his stomach feels better for a time.

But as the stomach gets more inflamed it demands more liquor, and the frequency and quantity are increased, until an appetite is formed. Now the body is constantly craving more fluid to quiet the disturbed stomach and other organs. The man who is in the habit of using alcoholic drinks is a sick man, and should be under the care of a physician, that his diseased body may be restored to health, — if indeed it be not already too late to have it restored.

The changes and effects we have described, come from the continued use of even small doses of alcohol. When large doses are taken within a comparatively short time, several results may follow. The most general one is acute inflammation of the stomach. The effects of this acute attack may disappear after a few days, provided no more alcohol be taken.

The prolonged use of alcoholic liquors causes marked changes in the structure of the stomach. The blood vessels become permanently enlarged, and the tissue is so changed that in places it dies, or breaks down, forming ulcers. These ulcers may be minute and widely scattered, or one or more large ulcers may be formed. The glands become much reduced in size, and the gastric juice is weakened and unable to do its work. Thus the most severe forms of disease, with distressing pain and great loss of strength, directly follow. This condition of things may go still further, until the stomach will not retain any food given it, and a lingering and painful illness at last terminates in death. When a person has reached these last stages of chronic alcohol poisoning, he craves more alcohol or opium to quiet the burning pains, and there seems no relief from the terrible bondage.

When we consider that one glass of liquor creates a desire for another, and when we understand to what all this may lead, we are prepared to agree that the best way to deal with such a drug is to let it alone.

The Effect on the Liver. It is probable that nearly all the alcohol taken into the stomach is there absorbed, and that but very little, if any, passes out into the intestines. If it is taken up by the·blood vessels of the stomach, the first organ to which it is carried is the liver. Here it does immense damage. The liver, like all other organs, is made up of cells. These cells should have but very little, if any, fat in them. If their material is changed into fat, they can no longer either secrete bile, or store up the sugar for use in the body. But alcoholic drinks cause the liver to become large and fatty. If their use be continued, at a later stage the liver becomes smaller and harder than in health. This hard, small liver is so characteristic that it has been given a distinct name by medical men,— it is called "the drunkard's liver." The liver is doubtless the first organ to suffer from the use of alcoholic drinks. It undergoes an actual change in its structure, and brings about a general disorder of the whole system as a result. The liver is made to suffer in other ways as well; but these two results are sufficient to show the ill effects of this drug, — first, an enlargement and fatty change; and second, after prolonged use, a shrinking and hardening of its structure.

Effect of Tobacco on Digestion. The effect of tobacco on digestion is largely of a secondary nature. It first affects the digestion of those who chew, because the salivary glands are so continuously overworked that when the saliva is most needed, at meal-time, only a scanty amount is furnished. Its more severe effects are shown through the nervous system, causing a particular kind of indigestion, called nervous dyspepsia.

The Tobacco Cancer. It is well enough to call attention, in this connection, to another bad effect from the use of tobacco. We refer to the " tobacco cancer," or " smoker's cancer." The irritation caused by having the poison of the tobacco so constantly on the surface of the lip may give rise to a small ulcer, which develops into a cancer.

No Use for Tobacco. While the evil effects of tobacco are not equally manifest in all its users, it is evident that the habitual use of such a poison must sooner or later do harm. Sometimes this harm affects others. The children of chewers and smokers often inherit weak nerves and impaired vital force from the tobacco habits of their parents. To the growing boy or the rapidly developing young man, tobacco is most certainly injurious. The younger the person who uses it, the more harmful it is. One of the worst things that can be said about this useless weed is that it is a narcotic, and thus deadens the finer sensibilities; as a rule it is found to make the person who uses it thoughtless of the comfort and convenience of those about him. To satisfy his own desires he will smoke wherever and whenever he chooses. Everybody should breathe pure air. The smoker who befouls the air that others must breathe with nicotine fumes is trespassing on the rights of others.

Opium. This drug will completely check digestion It takes away the appetite, checks the flow of the gastric juice, and deranges the whole digestive apparatus.

Tea and Coffee. When taken very strong and in large quantities these beverages are quite sure to interfere with digestion.

QUESTIONS.

1. Is pure alcohol generally used?
2. How is it usually taken?
3. What is the first effect of strong drink on the stomach?
4. What important result follows this?
5. What serious changes occur in the mucous membrane?
6. How does this affect the gastric juice?
7. What disease follows? To what does it give rise?
8. Why is the drinker deceived into taking more of the drink that has caused this trouble?
9. How is an appetite for alcohol formed?
10. Why does the man now crave strong drink?
11. What is the general effect of small doses? Of large doses? Of prolonged use?
12. What is said about the formation of ulcers?
13. What is the effect of alcohol on the glands?
14. Where is most of the alcohol absorbed?
15. Which is the first organ it reaches?
16. How does alcohol affect the liver?
17. Give the two most marked effects.
18. How does tobacco affect digestion?
19. How is the tobacco cancer formed?
20. How does tobacco affect others beside the user, and show that it deadens the sensibilities?

CHAPTER XVI.

ABSORPTION.

What has thus far been done. The saliva has changed some of the starchy foods into sugar ; the gastric juice has digested the lean meats, egg, etc.; while the pancreatic juice has digested the fats, changed all the rest of the starches into sugar, and finished the work of the stomach, if necessary, and the bile has flowed into the intestine to give its aid. All this work has been done in order that the foods which cannot be absorbed in their natural state may be changed to a liquid form, and thus taken up by two kinds of vessels,— the blood vessels and the lymph vessels.

Absorption from the Stomach. The water that is taken into the system is generally absorbed by the blood vessels of the stomach, and thus it enters at once into the general blood current. There is some slight absorption of digested food, but this occurs principally in the small intestine.

Villi of the Small Intestine. Hanging down from the inner walls of the small intestine are minute projections, like fingers ; these are called villi. It is estimated that there are over four millions of them in the human body.

Figure 31, B, shows a cross section of the small intestine, with a number of these little fingers, or villi, hanging into the central cavity. It also shows the two kinds of vessels. Let us place one of these villi under the microscope. We now see that a layer of cells

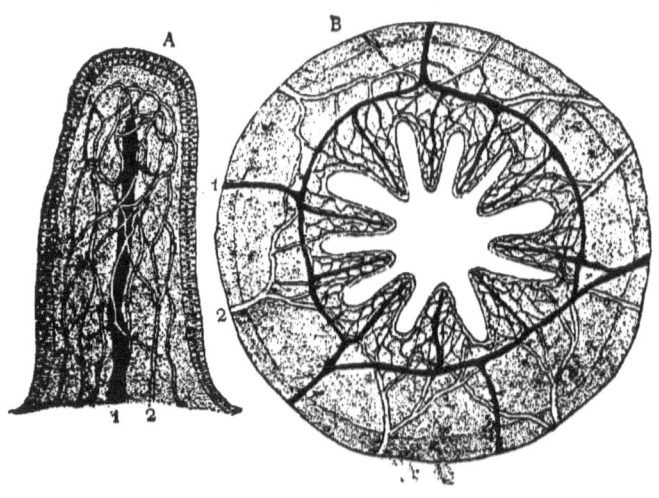

FIG. 31. B, a cross section of the small intestine. A, one of the villi of B, highly magnified: (1) the lacteals, or lymphatics ; (2) the blood vessels.

surrounds it, while its inside consists almost entirely of the two kinds of vessels. Directly in its centre is the lacteal, or lymph vessel. There is also a large number of minute blood vessels. What are these villi for ? What is the single lacteal for ? Of what use are so many little blood vessels ?

How the Foods are absorbed. As the fine rootlets of a plant soak up nourishment from the ground, so these villi take up the digested food. It would not be far from correct to say that the digested food soaks through the thin walls of the villi as water soaks through a cloth. Once in these little vessels, the absorbed food flows into larger vessels, as shown in Fig. 31, at the right, and these will soon unite together in one or two still larger vessels. Rivers are formed on the same plan. Small streams flow from many directions; these flow into each other to make larger ones; and finally these unite together to make one broad stream.

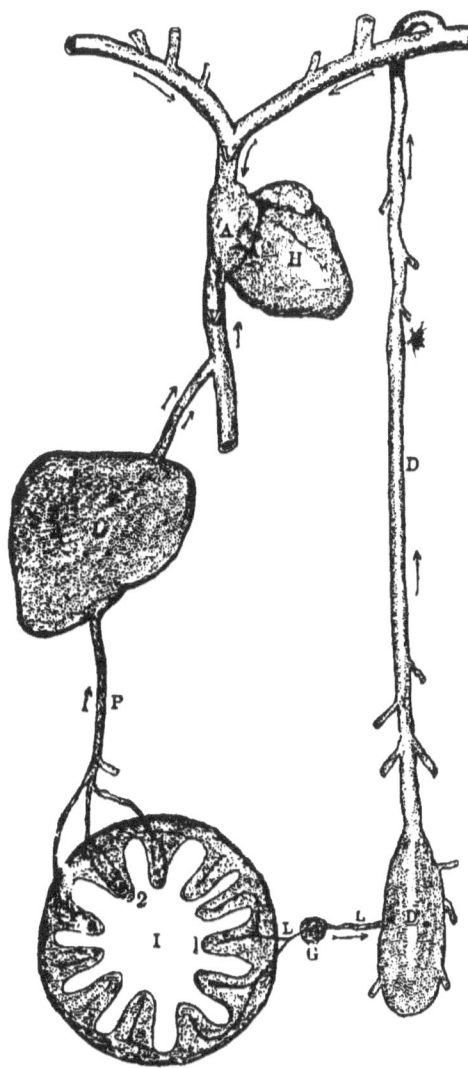

Although the blood vessels and lacteals absorb the digested foods yet the work of the lacteals, in the centre of the villi, is largely to take up the

FIG. 32. I, intestine : (1) villi with central lacteals; (2) villi with blood vessels. L, lymphatic or lacteal vessels. G, lymphatic gland. D, thoracic duct. P, portal vein. L, liver, at the left of the figure. V, vein. H, heart. A, right auricle of heart.

digested fats, while the blood vessels take up principally the other digested foods.

The Portal Vein. The blood vessels of the villi soon unite with each other and with those of the stomach to make a large vein, called the portal vein. This goes directly from the intestine to the liver. Fig. 32 shows where this vein begins, and where it ends. Its duty is to carry the digested foods to the liver. From this organ they will be conveyed into a large vein in the direction of the arrows of the figure. Thus the blood, with its digested foods, gets into the right side of the heart, and from the heart it will soon reach all parts of the body.

The Thoracic Duct. The lacteals, or lymphatics, that are in the centre of the villi soon unite to make a large vessel, called the thoracic duct. This duct is as large as an ordinary slate pencil, and it lies in front of the spinal column. It ends above by emptying into the large vein just beneath the left collar bone, and this vein goes directly to the right side of the heart.

The Lymphatic Glands. There are small glands situated all along the course of the lymphatic vessels. These are the lymphatic glands.

The Lymph Corpuscles. If some of the lymph from the thoracic duct be examined with the microscope, there will be found in it a vast number of minute bodies called lymph corpuscles. When they get into the blood they are called white blood corpuscles. These will be described later, when we study the blood.

How the Digested Foods enter the Circulation. Let us first examine the course of some of the foods digested in the stomach, taking broiled steak as an example.

Lean Meat. First, the beef is thoroughly chewed, or masticated : second, it is swallowed : third, it is digested by the gastric juice ; fourth, it is forced into the intestine by the contractions of the stomach ; fifth, it is absorbed from the intestine, principally by the blood vessels of the villi ; sixth, it is carried by these vessels into the portal vein ; seventh, it is carried by the portal vein into the liver ; eighth, it passes out of the liver through the large veins which carry it to the heart.

Starch. The starchy foods pass through the same channels, but the saliva and the pancreatic juices digest them, and they remain a longer time in the liver. stored up until they are needed.

Fat. We will next follow some of the fatty foods. First, they are masticated, if necessary ; second, they are swallowed ; third, they pass out of the stomach unchanged ; fourth, in the intestine they are mixed with the bile and the pancreatic juices, and so changed that they can be absorbed ; fifth, they are absorbed mainly by the lacteals ; sixth, the lacteals carry them through the lymphatic glands to the bottom of the thoracic duct : seventh, they pass up this duct and empty into the large vein which carries them directly to the heart. A study of Fig. 32 will make this clear. Begin the study with the intestine, I.

Fatty Foods. We shall suppose that the lacteals of the four villi at the right (1) are filled with the digested fatty food, called the chyle. The lacteals pass out of the intestine and unite with each other to form larger vessels, which pass through glands, as shown at G. These vessels empty into the lower end of the thoracic duct, D. This end is expanded into a sac. The chyle passes up the duct and empties into the large vein, which soon carries it to the heart.

Other Foods. Let us follow the course of such digested foods as meat and egg and the digested starches now changed to sugar. We shall suppose that the blood vessels of the villi, marked 2, are filled with the digested foods. These vessels soon unite to make the portal vein, P. This vein carries the food to the liver, L. In due time the liver will send it out through the veins in the direction of the arrows, thus reaching the heart. In this manner we are able to trace any food from its natural state until it gets into the blood.

The Lymphatics. Besides the blood vessels there are other fine vessels distributed all over the body, called lymphatics. These small vessels unite to form larger ones, which eventually empty into the blood vessels. The lymphatics collect the old, worn-out materials from all parts of the body and carry them to the blood vessels, from which they are taken by organs made especially for such work, as the kidneys and other glands.

CHAPTER XVII.

THE BLOOD.

Amount of Blood in the body. About one twelfth of the weight of the body is composed of blood.

It is very generally distributed. Blood flows from any part of the body, when the skin is cut through, though there are certain portions of the body in which we cannot find blood; namely, the hard parts of teeth, the hair, the nails, the outer layer of the skin, some parts of the eye, and most of the cartilages. These are nourished by fluids which soak through from the blood vessels.

Composition of Blood. Blood appears to the unaided sight like any other red liquid ; but the microscope shows that there are two parts to it, — first, a watery fluid, called the plasma, and second, some minute bodies, known as the blood corpuscles. The liquid plasma looks like so much water, but we know there are many important substances dissolved in it which we cannot see and which can be found only by the chemist. The blood corpuscles may be seen and studied with a powerful microscope. We thus learn that there are two kinds, known as the white and the red corpuscles.

White Corpuscles. The white corpuscle, as its name tells us, is without color. When persons become weak and pale, the microscope shows that their blood has in it too many of the white corpuscles, and not enough of the red. There ought to be only one white corpuscle to about three hundred red.

Red Corpuscles. The red corpuscles of man are circular bodies that are slightly hollowed towards the centre. In a great many animals they are of this same shape, but in birds, fishes, snakes, and some other animals, they are oval. Fig. 33 shows the oval corpuscles

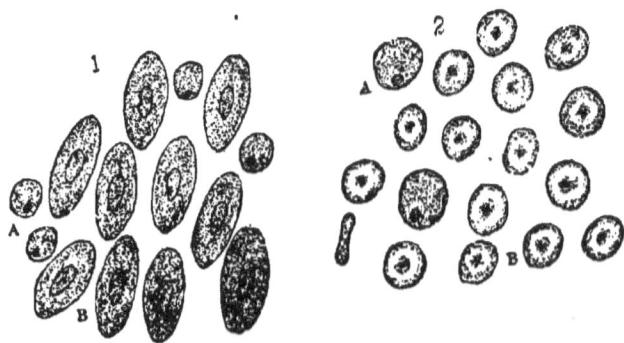

FIG. 33. (1) Frog's blood; (2) Human blood. A, the white corpuscles; B, the red corpuscles.

of frog's blood, and the circular ones of human blood. One red corpuscle of man's blood is represented as seen on the edge, which shows how it is hollowed towards the centre. These corpuscles are very small. If placed side by side in a straight line, it would take over three thousand of them to cover a single inch in length.

The Number of Red Corpuscles. It is impossible for us to realize how many red corpuscles there are in the body ; in a small drop of blood there are as many as five millions.

Their Use in the Body. The red corpuscles may be compared to little circular boats floating in the water of the blood. They go to the lungs, where they get very near the air, and take from it all the oxygen they can carry. Then they hasten away to some distant part of the body where some tissue needs it. To such tissue the corpuscles give up their oxygen while they are going through the capillaries ; then they hasten back to the lungs for another load. For this reason, the red corpuscles are called oxygen carriers.

Arterial Blood. When these corpuscles have a great deal of oxygen in them, or when they have just left the lungs, they are bright in color, and they make the whole blood appear bright scarlet. This bright-colored blood is found in the arteries, hence it is called arterial blood.

The Pulmonary Veins. There is one place where this bright blood is found in the veins ; namely, in the vessels that carry the blood from the lungs to the left side of the heart. These vessels are called the pulmonary veins.

Venous Blood. After the corpuscles have passed through the capillaries, and have there given up their oxygen to the tissues, they become darker, and as a result the whole blood looks darker. Dark blood is found in the blood vessels which extend from the capillaries through the body and back to the lungs.

The Pulmonary Artery. The pulmonary artery carries blood from the right side of the heart to the lungs. It must, therefore, contain venous blood. We say that the arteries contain arterial blood, and the veins venous blood; but to this rule there are the two exceptions we have just given, in the cases of the pulmonary artery and the pulmonary veins.

The Air. The air we breathe consists principally of two gases, — oxygen and nitrogen.

Oxygen and Carbonic Acid Gas. All parts of the body need oxygen. If we did not have it we should die in a few moments. It is necessary that we should have it in large quantities; for this reason we are more healthy if we always breathe fresh and pure air.

There is another gas we must study in this connection. Carbonic acid gas is unlike oxygen, for no animal will live if placed in it. It is a deadly poison to all animal life. After the tissues have taken up the oxygen from the blood they give back to it this carbonic acid gas.

Difference between Arterial and Venous Blood. The principal differences between arterial and venous blood are these : —

Arterial blood has more oxygen than has venous blood.

Venous blood has more carbonic acid than has arterial blood.

Arterial blood parts with its oxygen in the capillaries.

Venous blood parts with its carbonic acid in the lungs.

Arterial blood is of a bright scarlet color.

Venous blood is of a darker, nearly purple color.

The Clotting of Blood. After blood has flowed out of the blood vessels for a short time, it thickens into a jelly-like mass. This is called the clotting, or coagulation, of the blood. Blood does not clot while it is in the vessels unless there is some disease. Sometimes a blood vessel breaks in the brain, and a small amount of blood escapes into the substance of the brain. Then it clots, and this clot causes the disease known as apoplexy. It is because blood clots that we do not bleed to death when we cut through the skin, or in any way cause the blood to flow. The little clot that is formed at the opening of the blood vessel closes it and the flow ceases. If the blood flows slowly, it will clot more easily. So when we are wounded in any way, we should by checking its flow for a short time help the blood to clot. This may be done by pressing on the wounded spot and by keeping it very quiet, so that after the clot has formed in the end of the injured vessel it will not be disturbed.

The Blood in case of Injury. When one is injured, we can tell whether the blood is from an artery or a

vein, not only by its color, but also by the way it flows.
If it comes from an artery, it flows in jerks; we say the
blood spurts. If it comes from a vein, the flow will be
in a steady stream. It is more dangerous to have an
artery injured than a vein, as the flow from an artery
is not so easily stopped.

ALCOHOL AND THE BLOOD.

Surgeons tell us that wounds do not heal as readily
on those who are in the habit of using alcoholic drinks
as they do on those who never use them. This is because the alcohol has affected the blood in some way,
so that an injury is more likely to be followed by
inflammation.

QUESTIONS.

1. State the amount of blood in the body.
2. Describe its general distribution.
3. What does the microscope show in blood?
4. What is said about the white corpuscles?
5. Describe the red corpuscles.
6. Give an idea about the number of red corpuscles.
7. Of what use are they? What are they called?
8. What makes blood a bright scarlet? What is this blood called?
9. Where is the bright, arterial blood found in veins?
10. What makes blood look dark, and where is the dark blood found?
11. Where is the pulmonary artery?
12. How important is oxygen?
13. Is carbonic acid gas important to life?
14. Give a few differences between arterial and venous blood.
15. Tell some facts about the clotting of blood.
16. What is said of the healing of wounds in those who use alcoholic
    drinks?

CHAPTER XVIII.

CIRCULATION.

To understand how the blood is carried to all parts of the body we must study the heart, the arteries, the capillaries, and the veins.

The Heart. The heart is a large involuntary muscle. In shape it is something like a pear, with the small end down and to the left. It is situated in the chest with the lungs. By reference to Fig. 29, it will be seen that the heart and lungs nearly fill the thoracic cavity, and that they are separated from the organs in the abdomen by the thin wall of muscle called the diaphragm. The heart is enclosed in a sac, the lower part of which rests on this diaphragm.

The Position of the Heart. The heart is not all on the left side. The rule is that if we draw a line down the middle of the breast-bone, as shown in Fig. 34, the heart will extend about three inches to the left of the line, and one and a half inches to the right. The base, or larger part, is up as high as the third rib, and more of the base is on the right side than on

8

the left. The point of the heart extends well over the left side : it is there we can feel it beat. The illustration shows that this point is between the fifth and sixth ribs, and that the heart is placed obliquely in the chest.

The Cavities in the Heart. The heart is di-

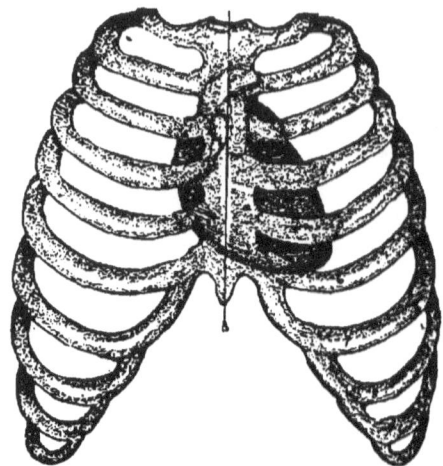

- FIG. 34.   The position of the heart.

vided lengthwise, by a firm wall, into two parts, so that there is no connection whatever between the two sides. The left side always contains arterial blood, and the right side venous blood. By reference to Fig. 35 it will be seen that the parts 3 and 4 together represent the right side, while the parts 6 and 7 together represent the left side. The wall between the two

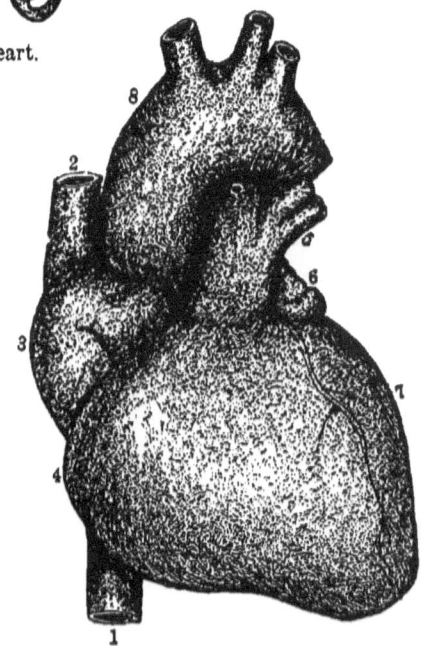

FIG. 35.   The heart, and the larger vessels at its base or upper part : (1) and (2) veins ; (3) right auricle ; (4) right ventricle; (5) pulmonary artery; (6) left auricle; (7) left ventricle; (8) aorta.

is shown by the location of a small blood vessel seen on the outside of the heart to the left of number 7. By examining each side a cross partition is seen, dividing it into two parts. Numbers 3 and 4 represent the two parts of the right side, while numbers 6 and 7 represent the two parts of the left side.

The wall that divides the heart crosswise is not a complete one, but has openings through it,

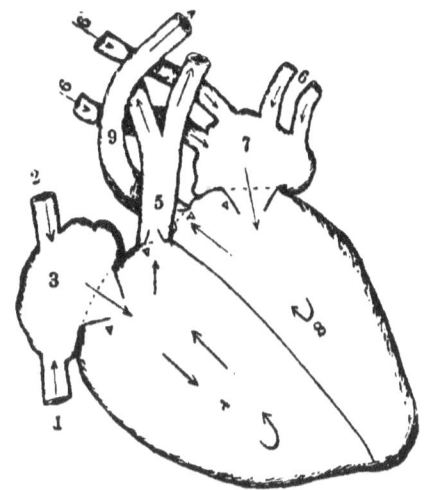

FIG. 36. Diagram illustrating the flow of blood through the heart : (1) and (2) large veins that bring the blood back to the right auricle ; (3) right auricle ; (4) right ventricle ; open valves between them ; (5) pulmonary artery, carrying blood to the lungs ; notice the open valves ; (6) pulmonary veins, bringing blood from the lungs : (7) left auricle ; (8) left ventricle ; notice the valves ; (9) the aorta ; notice the valves.

so that blood can pass from 3 to 4 on one side, and from 6 to 7 on the other. These openings are protected by doors which Nature has provided ; these doors are called valves. In conse-

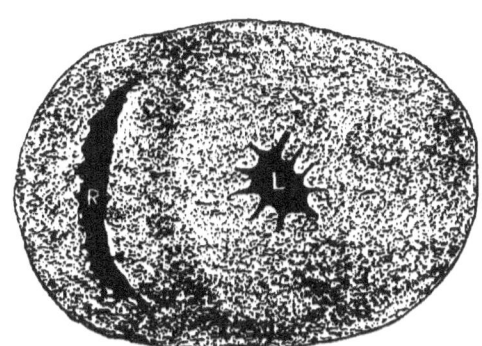

FIG. 37. A cross section of the ventricles of the heart.

quence of these walls, or partitions, there are four cavities in the heart. The two upper cavities are called the auricles. They are so named because they look like ears, for the word auricle means ear. The four cavities are as follows: the right auricle, the right ventricle, the left auricle, and the left ventricle.

The Contraction of the Heart. Each ventricle holds from four to six ounces of blood. When the heart contracts, it makes its cavities smaller, thus driving or pushing the blood that is in it into the large channels which lead from its upper part.

The Course of the Blood through the Heart. Let us follow the circulation of the blood through the heart. Notice Fig. 36, and follow the numbers in order, also the arrows. Two large veins, 1 and 2, bring the blood from the body back to the right auricle. The right auricle, 3, contracts and pushes the blood into the right ventricle, 4. The right ventricle contracts and pushes the blood into a large vessel, called the pulmonary artery, 5. This soon divides into two vessels, one going to each lung. After passing through the lungs the blood is brought to the left auricle, 7, by the pulmonary veins, 6. The left auricle contracts and pushes the blood into the left ventricle, 8. The left ventricle contracts and pushes the blood into a large artery, 9, which carries it to other vessels, as shown by Fig. 38. Briefly, the circulation follows this course: from large veins into right auricle; then through right ventricle; then through

the lungs: then to
the left auricle ; then
through left ventricle ;
then out to the body.

**Power of Contraction
greater on one Side than
the Other.** The right ventricle of the heart has
to contract with just
force enough to send
the blood to the lungs,
which are only a short
distance away ; but the
left ventricle has to contract with force enough
to send the blood to
distant parts of the
body. Hence the walls
of the left ventricle
are much thicker and
stronger than those of
the right. Fig. 37
shows a cross section
of the heart when it is
contracted. The left
ventricle is at the right
of the figure.

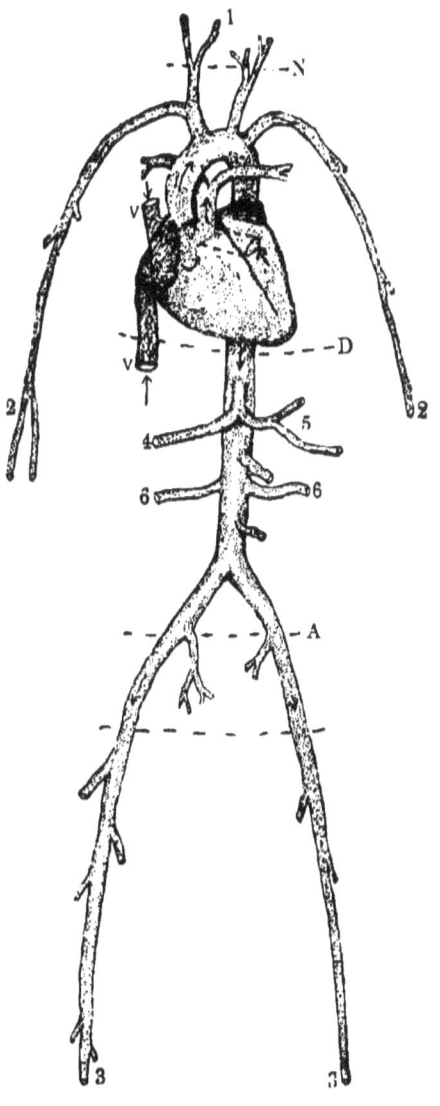

FIG. 38. The general plan of the circulation: N, the neck; D, the diaphragm; A,
the lower border of the abdominal cavity ;
V, veins returning the blood to the right
side of the heart. All other vessels represent arteries, which carry blood to
the following parts : (1) to the head; (2) to the arms; (3) to the legs; (4) to
the liver; (5) to the stomach, pancreas, and spleen : (6) to the kidneys.

The General Circulation. A study of Fig. 38 will show how all the large arteries that carry blood to the different parts of the body come from the artery that curves back from the upper part of the heart and passes down the central part of the body. In Fig. 35 this vessel (the aorta) is indicated by number 8, and in Fig. 36 at number 9.

The Valves in the Heart. By reference to Fig. 36 some valves will be seen. They are represented as doors that can open but one way; hence they will not let the blood flow in the wrong direction. If the blood should try to pass back, these doors would tightly close. All the doors are represented as open. These valves are seen between 3 and 4, between 4 and 5, between 7 and 8, and between 8 and 9.

The Heart rests. All the tissues of the body must have rest, so the heart has its time to rest; for after a part has contracted, there is a slight pause before it contracts again. This time seems short, yet when all these short periods of rest are put together they amount to between six and eight hours of each day.

The Heart works. The heart does an immense amount of work. Suppose the heart beats seventy times each minute : this would give over four thousand beats an hour, or nearly one hundred thousand a day. If all the work that the heart does in a single hour could be done at once, it would equal the force required to lift five tons of coal a foot from the ground.

How fast does the Heart beat? Some things will make the heart beat fast, and others will make it beat slowly. Sorrow and depression of spirits make it beat slowly. Excitement, as joy, anger, etc., make it beat fast; exercise makes it beat fast. It beats faster when we are standing than when we are sitting; faster when sitting than when lying; faster when awake than when asleep.

The number of beats is about ten more each minute in women than it is in men: in man it is between sixty and seventy a minute. Some persons have naturally either a slow or a quick pulse. The number is less in old age, while in young children it is as high as 120 to 140 a minute.

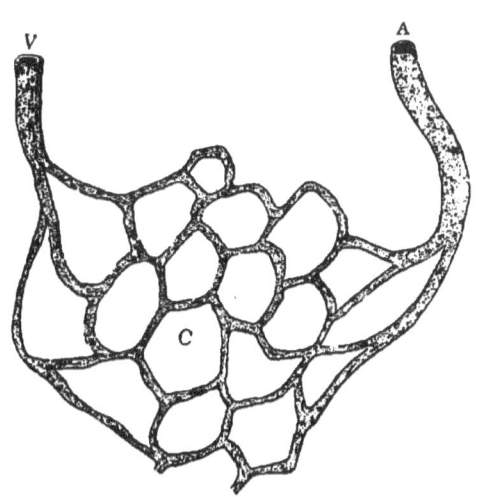

FIG. 39.   A, a small artery; C, capillaries; v, a small vein.

The Sounds of the Heart. Each time that the heart beats it makes two sounds. These can be distinctly heard if the ear be placed over the heart. One sound quickly follows the other, and then there is a period of silence. You will notice that these sounds are not alike. They are always of a certain character in health, so that the physician is able to tell. by listening to them. whether the heart is diseased or not.

The Pulse. The pulse at the wrist is caused by the sudden expansion of the artery. The heart pushes so much blood into the arteries that it makes them swell, or expand, at each beat.

Arteries, Veins, and Capillaries. The largest artery in the body is the one that comes from the left ventricle of the heart; it is called the aorta. Fig. 38 shows that not far from the heart the aorta gives off many branches. These branches divide again and again, until they become so small that they cannot be seen with the unaided eye, and a microscope is necessary to study them. These fine branches are called capillaries.

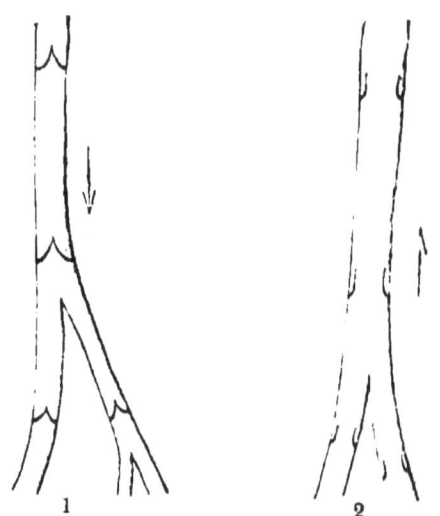

FIG. 40. (1) A vein, with the valves closed. The blood cannot flow in the direction of the arrow ; (2) a vein, with the valves open. Nothing prevents the blood from flowing in the direction of the arrow.

Fig. 39 illustrates the way in which a small artery divides to make the capillaries, and also how these capillaries unite to make a small vein. It is while the blood is in these capillaries that the tissues take from it the oxygen and other nourishment for their growth and repair.

Valves in the Veins. We found that there were valves in the heart to keep the blood from flowing in the wrong direction : and we find that there are many valves in the veins for the same purpose. Fig. 40 shows the arrangement of these valves. In the vein marked 2 the valves are open, and the blood is flowing in the direction of the arrows without anything to prevent it. But if the blood should attempt to flow in the opposite direction, as shown in the vein marked 1, then the valves would close and completely shut off the passage.

How rapid is the Circulation? The heart contracts with such force that the flow of blood in the large vessels near it is very rapid, but when the blood passes through the minute capillaries, the rapidity of its current is greatly diminished. The current in the veins is not so rapid as in the arteries, and is the slowest in the capillaries. The heart contracts with such force and frequency that a quantity of blood can leave it and go to the most distant parts of the body and get back again to its starting place in less than half a minute.

SUGGESTIONS TO TEACHERS

1. Procure at the market the heart of a calf or sheep. Preserve the large vessels at its base ; wash in water and wipe dry ; call attention to its shape ; hold it obliquely, with the base up, to show its position in the body ; notice the auricles, ventricles, and large vessels at its base.

2. Illustrate the circulation of blood through the heart, by pointing to each part in order ; thus, through right

auricle, then right ventricle, then to the lungs, back to the heart, through left auricle, then through left ventricle, and out through the aorta.

3. Cut the heart open transversely, about half-way back from the apex. This will appear as in Fig. 37. Call attention to the firm partition between the sides.

4. Cut the ventricles away up close to the auricles. Notice the white thin membranes that are in the openings between the auricles and ventricles: these are the valves.

QUESTIONS.

1. Describe the heart.
2. What is its position in the body?
3. How is the heart divided, lengthwise?
4. What blood is in the right side? The left?
5. How is each side divided, crosswise?
6. Is this wall a complete one?
7. How are the openings protected?
8. Name the four cavities of the heart.
9. What occurs when the heart contracts?
10. Can you give, briefly, the circulation through the heart?
11. Which side of the heart has the more work to do? Why?
12. Of what use are the valves in the heart?
13. Does the heart ever rest? How many hours each day?
14. Does the heart work? Illustrate how much.
15. What conditions make the heart beat faster?
16. Why does sorrow and depression of spirits injure the health?
17. How does the position of the body affect it?
18. How fast does the heart beat? Does age affect the number?
19. Does the heart make any sounds?
20. What causes the pulse?
21. Tell about the valves in the veins.
22. How rapid is the circulation?

CHAPTER XIX.

THE EFFECTS OF ALCOHOL AND TO-BACCO ON THE HEART AND THE CIR-CULATION.

On the Heart. Alcohol affects the heart through the nerves which control it. The heart is not controlled by an effort of the will; it keeps steadily at work whether we command it to do so or not. It is, however, controlled by nerves, which regulate its beat. By experiments on the lower animals it has been proved that there are nerves which act as breaks to the heart; they keep the heart from beating too fast, and thus hold it steadily to its work. If these nerves lose their control in any way, the heart will beat faster.

We shall learn in another chapter that one great effect of alcohol is to paralyze the nerves. Here, likewise, alcohol tends to benumb, or partially paralyze, the nerves that control the beating of the heart; and thus it beats more frequently. This increased beating means that the heart has just so much less time in which to rest. As it has to perform so much extra work, one result is that its walls are thickened and its cavities enlarged. A later effect is still more to be dreaded. We have already said that the tendency of alcoholic drinks is to change healthy

tissue into fatty tissue; and the heart forms no exception to this rule. Its strong, muscular fibres become
changed into fatty material, which has no power to
contract. The whole heart enlarges, and its muscle
feels soft and flabby. The process of fatty change
(degeneration) has gone on so long that this muscle
cannot contract with power enough to send the blood
completely around its circuit, and dropsy, difficulty in
breathing, and other ailments surely follow. At last the
muscle becomes so weak that it can no longer contract,
and the heart suddenly fails to beat. There are other
causes of a fatty heart, but medical men everywhere
recognize the use of alcohol as the most frequent cause
of this incurable disease.

On the Arteries. The tissues that compose the walls of
the arteries also are liable to undergo this same fatty
change from the use of alcoholic liquors. As a result
they are weakened, and thus are liable to rupture. Rupture of an artery in the brain is the cause of the disease
known as apoplexy.

On the Smaller Vessels. There are nerves in our body
whose only function is to keep the walls of the small
blood vessels contracted to a certain size, so that they
will be more firm and not so liable to get too full
of blood. Alcohol paralyzes these nerves, and they
lose their power over the small vessels. This may
be only temporary, if a single dose of alcohol be
taken; but if the doses be repeated for some time,
this condition becomes permanent, so that entire free-

dom from alcohol will not restore the original power
to the nerves.

What does the red nose of the confirmed drinker
indicate? What do the red eyes and red cheeks indicate? They show that there is paralysis of some of the
nerves. They show that the nerves which should hold
the walls of the blood vessels firm have become power-
less, and that the vessels have become distended and
filled too much with blood. Redness of the eyes, nose,
and cheeks, denotes that the nerves must be greatly
affected by this poison.

The heart and the blood vessels feel the effects of this
harmful influence on the nervous system, and they show
it by the irregular beating of the former, and the dila-
tation of the latter. The red face and eyes of the
drinker tell a terrible tale, and should be an awful
warning to their owner that the most vital tissues of
his system, the nerves, are becoming profoundly affected. Many would stop their downward course if
they could. But the heart misses its excitant, the stomach demands more fluid to quiet its burnings; and
the nerves awake from their sleep and declare by their
aches and pains that things are going wrong. Something must be done. What shall it be?

The true physician steps in, and declares that the
patient is ill and needs careful attention. He pre-
scribes out-door exercise, treats the inflamed stomach,
allows only a limited diet until the liver recovers from
its abuse, and strongly advises that new friends and
associates be formed, and that no alcoholic liquors ever

be used again. If the power to resist temptation is sufficiently strong the weak body slowly recovers, — a happy result, which one could wish might always follow; but in too many cases the old appetite is stronger than the influence of the physician.

TOBACCO.

The active principle of tobacco, nicotine, will, if taken in sufficient quantity, completely paralyze the heart. When the fumes of tobacco are inhaled, only a very small amount of this poison enters the system. Still, even this small amount, when taken steadily for a long time, is liable to cause palpitation of the heart, as well as severe pain. Under the influence of nicotine, the beat of the heart becomes unsteady and irregular; and this irregularity follows so frequently the use of tobacco that physicians call this form of cardiac disease the "tobacco heart." So far as is known, the tissues of the heart are not changed by this drug; but it so affects the nerves that the heart does not keep steadily and regularly at work.

A medical writer has lately said that the pulse of persons using tobacco is from ten to fifteen beats per minute faster than is normal.

It is not necessary, in order to get the full effects of tobacco on the heart, that strong cigars be used. Some of the most severe cases result from the smoking of cigarettes. Palpitation of the heart is altogether too common among our young men, and the cigarette is largely responsible for the trouble.

CHAPTER XX.

RESPIRATION.

The Larynx. After the air has passed through the nose it reaches the larynx. This is sometimes called the voice-box, because it contains some membranes that are used in producing sounds. The expansion on the front of the larynx is commonly known as Adam's apple.

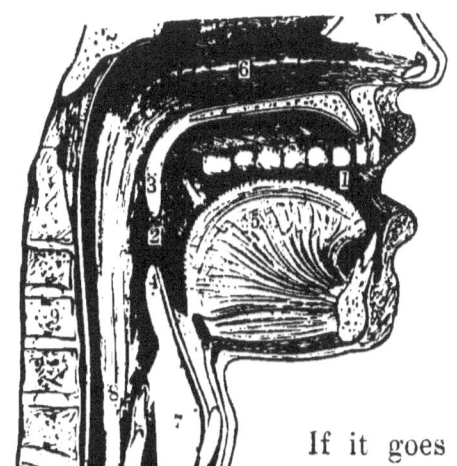

The Epiglottis. When food is swallowed, it passes down the œsophagus, or gullet. If it goes the wrong way, and passes down the larynx, it causes severe coughing. As these two canals lie side by side, as shown in Figs. 41 and 42, how is it that the food will pass down one, and the air pass down the other? It is because there is a little valve

FIG. 41. View of the inside of the nose, mouth, etc.: (1) the mouth; (2) position of the tonsils; (3) the uvula, commonly called the palate; (4) the epiglottis; (5) the tongue; (6) the nasal passage; (7) the larynx; (8) the pharynx.

at the top of the larynx. This shuts tightly over the larynx whenever anything is swallowed, but always remains open when we are not swallowing. This valve is called the epiglottis.

The Trachea. Just below the larynx is the trachea. which consists of rings of cartilage. These can be felt plainly in front of the throat. At its base the trachea divides into two branches, called the bronchi, one branch going to each lung. After entering the lungs, each branch divides again and again, until its branches are so small that it requires a microscope to see them. At the end of each little branch there is a collection of minute sacs, called the air cells.

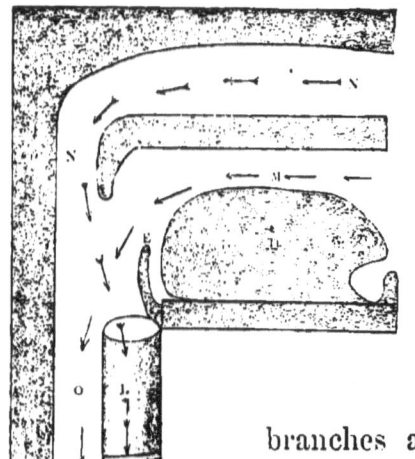

FIG. 42. A diagram illustrating the position of the epiglottis (F). (N) Nasal passage; (L) larynx; (M) mouth; (O) œsophagus; (T) tongue.

The Pleura. The inner walls of the chest are lined with a double membrane like a sac, called the pleura. This pours out a fluid which keeps its surfaces moist, so that when the lungs move against the walls of the chest they can do so easily and without pain. An inflammation of this membrane is called pleurisy.

The Lungs. There are two lungs, one in each side of the chest. They contract and expand; this is due to a

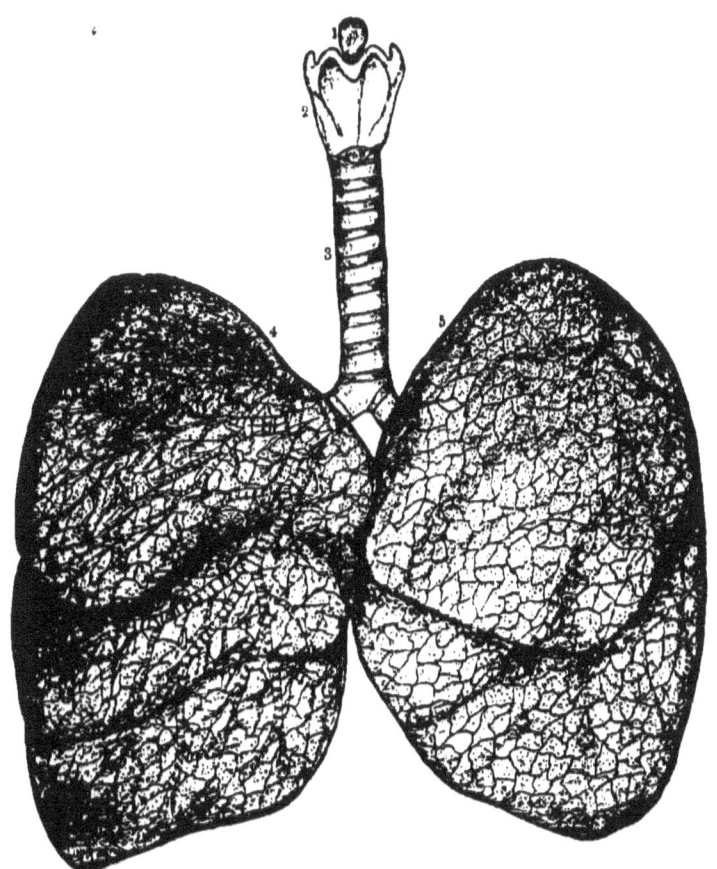

FIG. 43. The breathing organs: (1) the epiglottis; (2) the larynx, or voice-box; (3) the trachea, or wind-pipe; (4) the right lung; (5) the left lung.

tissue in them, which is something like rubber, and is known as elastic tissue. In its natural condition it is

9

contracted ; but when some force is applied, it stretches out like a piece of rubber. Filling the lungs with air stretches this tissue ; but it immediately contracts again, forcing the air out. From this we see that if we fill the lungs they will empty themselves.

How we Breathe. Why is it that air enters the lungs ? If you will watch the working of a pair of bellows, you will have the explanation. When the handles are separated the air rushes in. Why ? Because the air is pressing in every direction, and when the inside of the bellows is made larger by separating the handles, the air rushes in to fill the extra space.

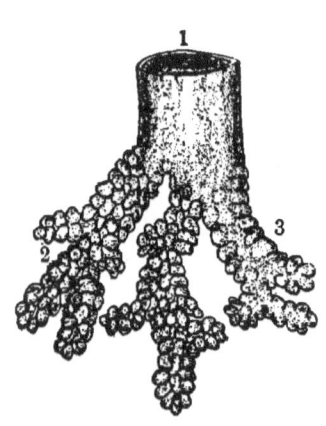

The chest is a tight box, with only one opening at the top, — the larynx. Suppose we suddenly make this box larger ; then more air rushes in through the open-

Fig. 44. (1) the end of a small bronchial tube; (2) air cells; (3) some of the air cells cut open, showing free passage to them from the bronchial tube.

ing. How is the chest made larger ? The diaphragm moves down, making the chest larger in that direction, while the ribs move and make the front and sides of the chest swell out. Then the air rushes in to fill the extra space.

Inspiration. Taking in the air is called inspiration.

Expiration. Breathing out the air is called expiration.

Respiration. One inspiration and one expiration, taken together, are called a respiration. There are from fifteen to eighteen respirations a minute.

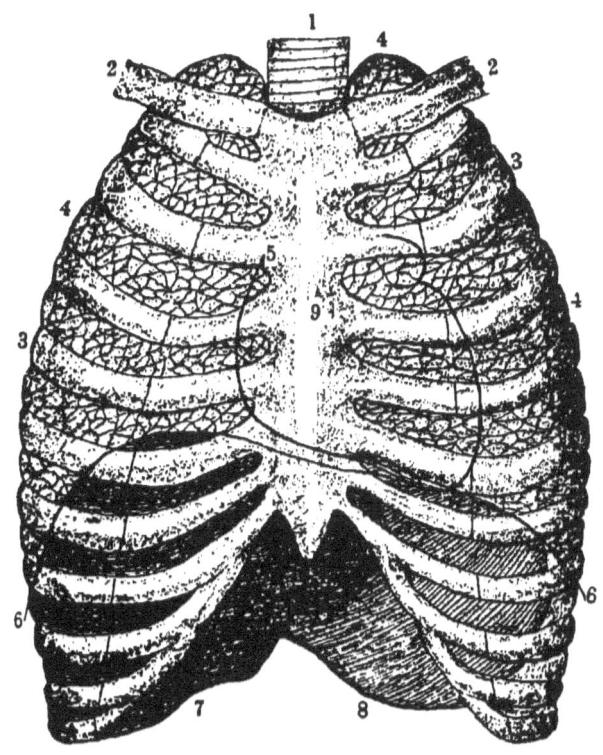

Fig. 45. This cut shows the position of the lungs and their relation to certain organs : (1) the trachea; (2) the collar bone; (3) the ribs; (4) the lungs: (5) the dark curved line shows the position of the heart; (6) the diaphragm, extending in a curved direction from one figure to the other; (7) the liver; (8) the stomach; (9) the breastbone.

Causes of a Respiration. Inspiration is caused by the enlarging of the chest, as described above. Expiration is caused by the elastic tissue contracting to its natural condition, thus forcing the air out.

Voluntary or Involuntary? Breathing is partly under the control of the will, and partly not. We breathe when we are not thinking of it, and we can for a time breathe either more slowly or faster than usual. Ordinary breathing, however, is involuntary.

Sounds of the Chest. When the air rushes in and out of the lungs, it makes peculiar sounds. By listening to these sounds the physician is able to tell whether the lungs are healthy or not.

Blood Vessels of the Lungs. There are a great many blood vessels in the lungs, together with great numbers of capillaries on the walls of the air cells, so that the air and the blood come very near each other.

Why do we Breathe? We breathe in order to get the oxygen that is in the air into our bodies, and also to get the carbonic acid of the body into the air. How is this done ? The red corpuscles of the blood take the oxygen from the air and carry it to all parts of the body, and the blood brings back to the lungs and gives to the air the carbonic acid it has received from the tissues. The blood while in the lungs receives oxygen, and gives up carbonic-acid gas.

The Expired Air. Expired air contains this carbonic acid and two other ingredients coming from the body. These are a watery vapor and an animal substance. As a rule, we cannot see the watery vapor; but on a cold, frosty morning, it freezes on the beards of men, and we

all say we can see our own breath. If we breathe on
any polished surface, as a mirror, or even the window,
the vapor will show. From one half-pint to one pint
of water in this way is given off from the body each
day.

The animal substance consists of a minute quantity
of material that gives a peculiar odor to the breath
of some animals, as the cow. The odor is not noticed
in the expired air coming from the healthy human
body; yet it putrefies very quickly, and we notice
it at once if we enter a poorly ventilated room in
which a number of persons have remained for some
time.

Results of Breathing. By breathing, the blood gains
oxygen, and loses carbonic acid, watery vapor, and some
animal matter. The blood does not take out all the
oxygen from the air in the lungs; therefore it is possible to live a short time breathing the same air over
again: but soon there is not enough oxygen left in this
air to support life.

How to Breathe. We should breathe through the nose.
for the following good reasons: the air will enter in
small quantities; it will get warmed in cold weather;
it will be moistened by passing over the moist lining of
the nose, and dust will cling to this membrane and will
not reach the lungs. If one breathes through the mouth,
the throat becomes dry, and in the winter the cold air
is likely to chill the warm blood of the lungs.

How One Takes Cold. One of the easiest ways to take cold is to pass from a heated room to the open air and inhale the cold air through the mouth. Keep the mouth closed, and breathe through the nose.

Tight-Lacing. The waist cannot be bound tight without interfering with the free action of the chest, which cannot enlarge as much as it should, and, as a result, the blood will not get sufficient oxygen. Binding the body also interferes with the circulation of the blood, and with the proper working of all the organs thus compressed. Clothing may be made to fit the body closely and firmly without harm; but tight lacing is very injurious.

ALCOHOL AND TOBACCO AND THE LUNGS.

Alcohol. As the heart beats faster and weaker under the influence of alcohol, the lungs are apt to be continually over-filled with blood. This tends to produce frequent attacks of bronchitis, colds, and other lung troubles. In addition to this, the alcohol in the blood of the drinker irritates the delicate lining of the air sacs of the lungs, making them more liable to disease. Medical writers of late years have recognized a rapidly fatal disease of the lungs which has been named alcoholic consumption. A very noted London authority, Dr. Peacock, who was in charge of a large hospital for consumptives, says that alcohol is a frequent cause of some of the severest forms of lung disease.

Tobacco. Tobacco is especially injurious to the lungs and throat. By its use the latter is made dry, and the voice becomes husky from the irritation of the poison. The membrane lining the larynx and the bronchial tubes is also irritated, producing a dry, hacking cough. The smoker's sore throat is a very common affection, and can be cured only by giving up the habit altogether. Some smokers inhale the smoke, or draw it into their lungs. This must certainly prove injurious.

QUESTIONS.

1. Give another name for the larynx. Why so called?
2. Why does not food enter the larynx?
3. What is just below the larynx?
4. Where do the bronchi begin?
5. Where are the air cells?
6. What is the pleura? What is pleurisy?
7. Describe the lungs.
8. Describe the action of the elastic tissue.
9. Explain the working of a pair of bellows.
10. How is the chest made larger?
11. Why does the air rush in?
12. What is inspiration? Expiration?
13. Describe and explain a respiration.
14. Is breathing under the control of the will?
15. What makes the peculiar sounds in the lungs?
16. Any blood vessels in the lungs?
17. How do we get oxygen into the body?
18. What does the expired air contain?
19. Give the results of breathing.
20. How does alcohol affect the lungs?
21. Is tobacco injurious to the lungs?

SUGGESTIONS TO TEACHERS.

1. Procure at the market the lungs of a sheep. Wash in water and wipe dry. Call attention to the rings of the trachea. Notice how the trachea divides before entering the lungs.

2. Put a tube of any kind in the trachea. Tie the trachea tightly about it. Put the mouth to the tube and force air through it, and the lungs will slightly expand; take away the mouth, and the lungs will collapse at once. The nozzle of a pair of bellows may be inserted in the tube, if preferred, and the lungs inflated by working the handles.

3. Cut off a small piece of the lung, and throw it into water. You notice that it does not sink, — which shows that all the air has not escaped. Press the piece in the hand, and try again. Still it will not sink. We cannot breathe out all the air that is in the lungs.

4. After an ordinary expiration let the pupil make an extra effort and breathe out more air. This proves that we do not exhale all the air in the lungs with each respiration.

5. Illustrate, by means of a pair of bellows, how the air will rush into a cavity when the cavity is made larger. The raising of one handle of the bellows corresponds to the lowering of the diaphragm and the swelling out of the chest.

6. Let the scholars illustrate inspiration, expiration, and a complete respiration.

7. Show that respiration is partly voluntary. Breathe fast, then slow, a few times.

8. Prove the presence of a watery vapor in expired air, by breathing on a mirror or on any polished surface.

CHAPTER XXI.

VENTILATION.

As we inhale about one pint of air each time that we breathe, and as the expired air has poisonous substances in it, we should be careful to have an abundance of fresh, pure air about us all the time.

How to obtain Pure Air. Cold air is not necessarily pure air, neither is a current of air always pure. Our rooms should be so arranged that there is an unfailing supply of fresh, out-door air pouring into them, and we should take care that our sleeping rooms are supplied with a constant change of air. An open grate, or a register, or a stove may cause a sufficient current. The air may be easily changed by raising the lower sash of one window and lowering the upper sash of another.

Avoid Currents of Air. If a current of air — a draught, as it is called — be allowed to strike on some sensitive part of the body, as the back of the neck, it is very likely to cause a cold, or something more severe. All currents of air should be avoided, especially when the body is moist with perspiration.

Odors. It is a fact that one odor can cover up another without destroying it. The odor of flowers may cover up the odor of a poorly ventilated room. Air may be very poisonous, and yet be made agreeable to the sense of smell.

Deodorizers. Any substance that will replace the odor of another, and yet not destroy it, is called a deodorizer. People burn coffee and sugar to destroy some odor that they fear will produce disease ; but thereby they only cover the odor with one more powerful.

Disinfectants. Disinfectants actually destroy odors. These are largely used by physicians to destroy disease-germs and to remove odors that are not only offensive, but also injurious to health.

Absorbents. Whitewashing a room sweetens and purifies the air, because the lime that is used absorbs certain gases. Lime and charcoal are both good absorbents.

Contagion. There are some very poisonous substances that do not let their presence be known by any odor whatever. This is true of the germs in contagious diseases, as diphtheria and scarlet fever.

Persons suffering from any contagious disease should be placed at once in a well-ventilated room, and all who who have not had the disease should be kept from them.

The author knows two families in New York State, living very near each other, which were attacked by diphtheria. The parents of the first family took the first child who was stricken with the disease into a room in the second story of the house, to which none of the remaining five children were allowed to go. In fact, no one was allowed to visit that room except the doctor and the mother. On the fifth day the child died, but not one of the other children had the disease. In the other house there were nine children. The father said he did not believe that a little sore throat was catching. Even after one of the children was stricken with the disease, the father would not consent that the directions of the physician should be carried out, and the children played together as usual. As a consequence, all nine children had the disease, and eight of them died within two weeks.

Nature has certain laws, and if we disobey them our punishment is often swift, and always certain. It is our duty to find out what these laws are, and then obey them.

QUESTIONS.

1. Why is pure air necessary?
2. How should our rooms be arranged?
3. How can we plan to have a change of air?
4. What about currents of air?
5. What is a deodorizer?
6. What is a disinfectant?
7. Why does whitewashing a room purify it?
8. Do all poisonous substances have odors?
9. Name some that do not.

CHAPTER XXII.

SLEEP.

ALL animals that have a well-developed nervous system must, at times, lie down to rest and to sleep. It is necessary not only that the brain have this rest, but also that the whole body have a share in it.

We must Sleep. By drowsiness and weariness we are warned that the body needs time to repair itself; and although this warning may be disregarded for a time, yet if we are in health, sleep will overtake us, sooner or later, no matter where we are.

How long shall we Sleep? Some persons require much more sleep than others, consequently no rule can be given that is suitable to every one. Infants sleep a good part of the time. It is said that Napoleon slept but three or four hours of the twenty-four. In middle life the average person requires about eight hours of sleep; before that age the time should be longer. It is a symptom of approaching trouble when one cannot get sleep. If one remains awake many hours beyond the usual time for sleep, the whole system feels the need, and most seriously objects.

To promote Sleep. Out-door exercise during the day, early and light suppers, rest during the evening, and warm feet, will tend to promote sleep. But, with all these, if the mind be filled with grief or anxiety or heavy care, it will not let the body rest.

Early to Bed. The midnight lamp of the student should be thrown out of the window. Better work can be done in one hour in the morning than in two late at night. Students need much sleep. They do not need the morning nap, however, as much as they need the sleep that comes before midnight ; therefore, " early to bed and early to rise " is as good a motto for the student as for the farmer.

Plenty of sleep is one of the laws that must not be broken. The night should not be turned into day, nor the day into night, in order to please our fancy. He who persists in breaking Nature's laws will pay a severe penalty ; he will lose that which is more precious than gold,— a healthy body.

Lay aside your books, go to bed early, have fresh air in the room, and lie down to quiet and restful sleep. Awake with the rising of the sun, and while the brain is refreshed, impress it with the teachings of the books.

We have seen that opium, chloral, alcohol, and tobacco are narcotics. These are often taken in various forms to produce sleep, which may be called " narcotic sleep."

NARCOTICS AND SLEEP.

Natural sleep strengthens the will.

Narcotic sleep weakens the will.

Natural sleep strengthens the muscles.

Narcotic sleep makes the muscles tremulous.

Natural sleep strengthens the nerves.

Narcotic sleep weakens the nerves.

Natural sleep strengthens the digestive organs.

Narcotic sleep causes dyspepsia, furred tongue, nausea, loss of appetite, and jaundice.

Natural sleep rests the whole body.

Narcotic sleep causes dreams, and does not refresh the body.

Natural sleep builds up.

Narcotic sleep tears down, exhausts.

Natural sleep makes one awake cheerful and pleasant.

Narcotic sleep is a cause of melancholy, and frequently leads to suicide.

QUESTIONS.

1. Do all animals need sleep?
2. How are we warned that rest is needed?
3. Can we disregard this warning?
4. Do all persons require the same amount of sleep?
5. What will tend to promote sleep?
6. When is the best time to study?
7. Give a few differences between natural sleep and sleep caused by opium or alcohol.

CHAPTER XXIII.

THE KIDNEYS.

THE kidneys lie close to the back, one on each side of the spinal column (see Fig. 29). In shape they re-

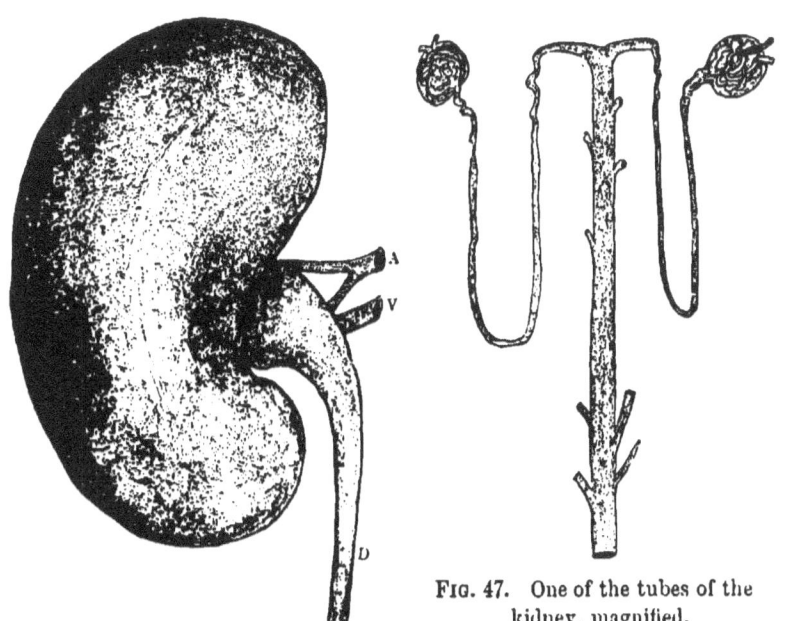

FIG. 47. One of the tubes of the kidney, magnified.

FIG. 46. A kidney : A, an artery; V, a vein; D, the duct that carries away the materials filtered from the blood.

semble a bean. They are composed of a number of minute tubes, which are so small that they can be seen only with a microscope.

The kidneys may be called the filters of the body. The blood is constantly flowing through them, and the tubes of the kidneys are steadily at work filtering from the blood certain materials that are very poisonous to the body. Should the kidneys fail to do their work, this poison would soon cause convulsions and death.

THE EFFECTS OF ALCOHOL ON THE KIDNEYS.

Alcohol has a severe effect on these organs, and gives rise to a serious disease, known as Bright's disease. There are other causes of this disease, but the use of alcoholic drinks is recognized as one of the most frequent of all.

QUESTIONS.

1. Where are the kidneys located?
2. Describe their shape and structure.
3. What work do the kidneys perform?
4. Is their work very important?
5. Tell how alcoholic liquors affect the kidneys?

CHAPTER XXIV.

THE SKIN.

THE skin is the protective covering of the body. Its outer part is composed of hard and dry cells, in which there are neither blood vessels nor nerves. This is called the cuticle. The cuticle may be pricked with a needle, causing neither pain nor the flow of blood. Just as soon, however, as the needle has reached the deeper part, called the true skin, it brings both pain and blood. It is possible to remove the cuticle by gently scraping the skin with a knife. The skin will then look red, and will be likely to bleed.

The True Skin. The true skin, called the cutis, is filled with nerves and blood vessels. It is impossible to put the point of a needle in it anywhere and not touch a nerve, giving pain, or a blood vessel bringing a drop of blood. There are small muscles which contract the skin, giving it the appearance known as goose pimples. Some of the muscles, as shown in Fig. 48, are fastened to the hairs in such a way that they can make the shorter ones stand more nearly erect. In the deeper parts of the skin are two kinds of glands, — the sweat glands, and the oil glands.

The Sweat Glands. A small magnifying glass will show the openings of the sweat glands on the ridges that appear so plainly on the ends of the fingers and the palms of the hands. These openings are close together, and look like

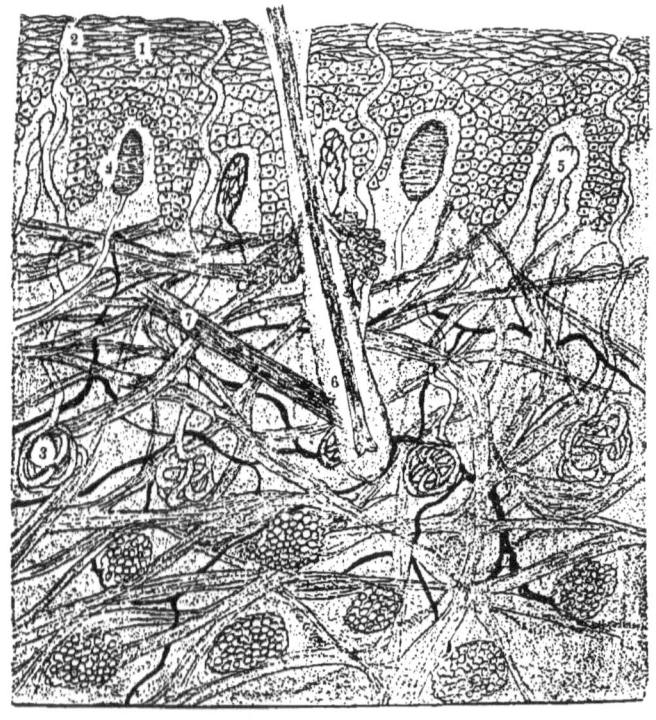

FIG. 48. Section of human skin, magnified: (1) the outer layer, the epidermis; (2) the duct of a sweat gland; (3) the gland itself ; (4) the ending of a nerve; for the sense of touch; (5) blood vessels; (6) a hair follicle; (7) muscle.

little pits, or depressions. They are the ends of tubes which go down into the skin about one fourth of an inch, and then coil up in a round mass like a ball. There are about three millions of these glands in the whole skin ; their function is to take water and some other substances from the blood and pour them on the surface of the skin, causing perspiration, or sweat. In fevers these glands are inactive, and the skin becomes hot and dry.

Two Kinds of Perspiration. Sensible perspiration is the secretion that accumulates on the surface in varying amounts. This is most marked when the body is active and when it is surrounded by warm air.

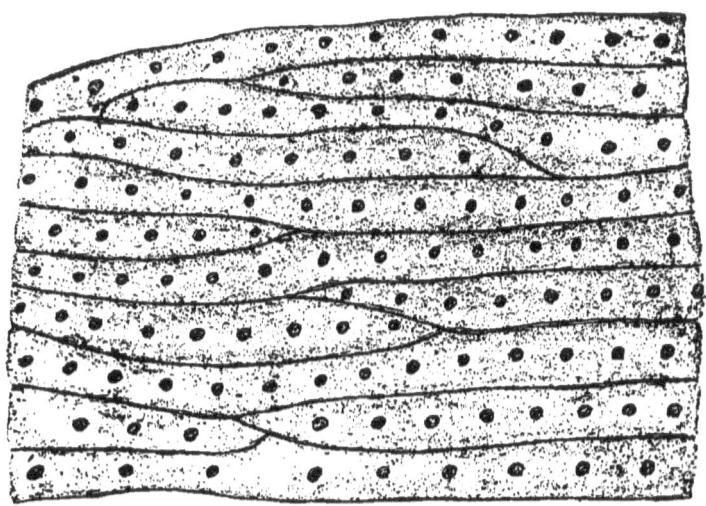

FIG. 49.   The surface of the skin magnified, showing the openings of the sweat glands.

Insensible perspiration is the secretion that is evaporated as soon as it appears on the surface. In health there is a constant secretion of the perspiration.

Checking the Perspiration. The perspiration carries off certain poisonous matters with it. If the body be suddenly cooled when it has been perspiring freely, the

work of carrying this matter off is at once thrown on
other organs, and this frequently causes disease in them.
As a result of this check, the kidneys frequently become
diseased, and the whole body may be thrown into a high
fever.   This is also one of the most common causes of a
cold.   After exercising, or whenever the body is perspiring freely, it is well to remember the following rules:
Do not get chilly; do not sit in a draught; do not remain in a cool room; do not drink too much cold water;
do not cool the body too quickly; let the body cool gradually, even throwing some light clothing over the shoulders while resting.   To perspire is good; to perspire freely
after exercise is good; but suddenly to check the perspiration, or to allow the body to cool rapidly, is positively
injurious, and may prove fatal.

The Oil Glands. The oil glands do not come to the
surface, as do the sweat-glands, but each empties into a
sac in which a hair rests ; so a hair has an oily substance
poured around it continually.   This is Nature's hair-
oil.   It makes the hair smooth and glossy, and, if the
scalp be in a healthy condition, it will furnish oil enough
to keep the hair soft and smooth.

The Hair. The hairs are placed obliquely in the skin,
as shown in Fig. 48.   Hair is not hollow, as many people
suppose, although in its centre there is a substance more
porous than the outside.

Hair grows from the bottom, from the bulb.   When one
hair is removed, another will grow in its place, provided

the deeper parts of the skin be in a healthy condition. There are many diseases of the hair, of which some cause it to fall out, others to shorten, and still others

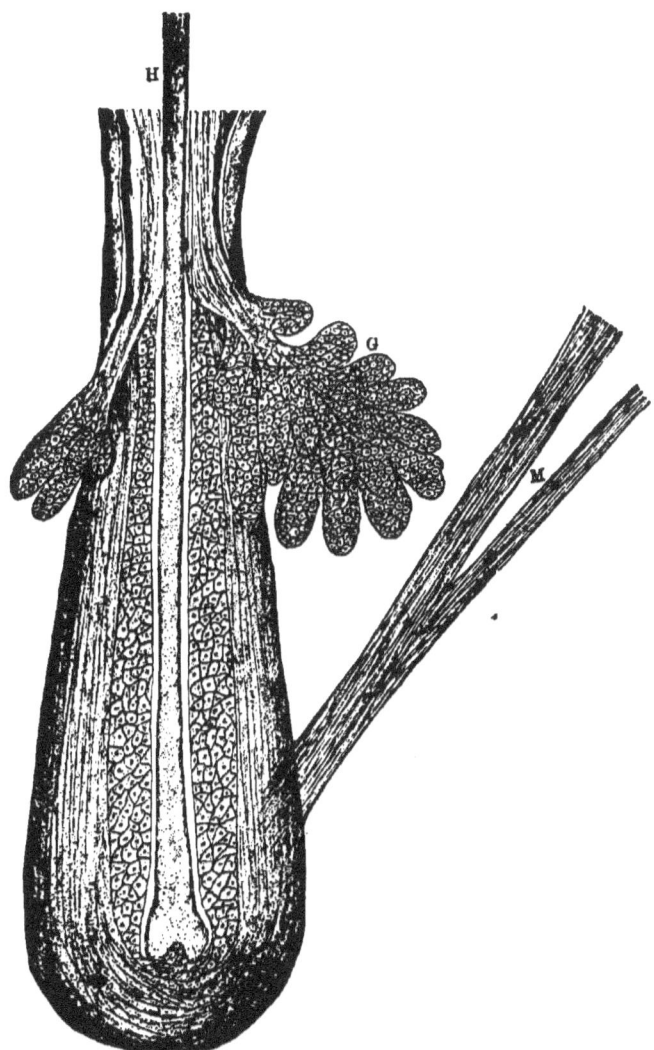

Fig. 50. A human hair in its sheath, or follicle, magnified : H, the hair; M, the muscle; G, the oil gland.

cause the hairs to split. When we grow old, Nature furnishes us with "a crown of beauty," not by giving us a dye to color the hair with, but by taking out the color of the hair and leaving it a beautiful gray or a pure white. In the majority of cases, though not all, it is possible to tell to what animal a hair belongs by its appearance under the microscope.

FIG. 51. Human hair as seen with a microscope.

FIG. 52. Cat hair as seen with a microscope.

The Use of the Skin. The skin is of great service to the body in many ways. It protects the delicate parts beneath it; it throws off waste products of the body; it regulates the temperature of the body; it contains the hair and the organs of touch.

Bathing. The whole body should be washed every day. Those who have long hair, however, would be wise not to wash the scalp daily, as if this were done it would take the hair too long to dry. It may seem strange to recommend that the body have a daily bath, when many people rarely bathe, and yet enjoy a fair degree of health. Still, this is no reason why our advice should not be heeded; for it is a fact that the skin has a certain amount of work to do, and this it cannot do well if the sweat glands do not properly act. It is evident that they cannot act if their openings are clogged up. If the skin cannot do its work properly, this work is left for the lungs and kidneys to perform. A neglect of the skin must make more work for the lungs, and especially for the kidneys. Only the purest and best soaps should be used on the body.

Why bathe ? We should bathe for the following reasons : —

Personal cleanliness is a duty.

Bathing removes the dried products of perspiration.

It makes the skin soft and smooth.

It removes dirt that has accidentally been put upon it, and accompanying odors.

Frequent bathing acts as a preventive to taking cold.

It keeps the skin active, and thus relieves the lungs and kidneys.

It enables the skin to act in a healthy manner, and thus removes one cause of fevers and skin diseases.

It causes a healthy exercise, and thus promotes all the functions of the body.

We should not bathe, —

When fatigued ;

Just before or just after a hearty meal ;

In too cold or too hot water ;
When the body is perspiring freely ;
In cold water, if we feel chilly before entering.

Hot baths are not so invigorating as cold ones. The body should be thoroughly dried, and rubbed briskly after each bath.

A Good Complexion. No one who has a sickly body should expect to have a fine complexion. That this may be secured the whole system must be in a healthy condition, and the skin must be active and strong. Cosmetics, powders, and hair dyes seriously injure the complexion. A healthy body has no more need of face-powders than it has of doctor's powders. But if the former are used the latter will quite likely be needed.

QUESTIONS.

1. Give a description of the skin.
2. What is found in the true skin ? What glands?
3. What does a magnifying glass show on the skin?
4. What are these pits ? Are there many of them?
5. Describe the kinds of perspiration.
6. What is the danger in checking the perspiration ?
7. How may we avoid a cold after exercising ?
8. Describe the oil glands. What is their use?
9. Give some facts about the hair.
10. Give some of the uses of the skin.
11. State a few reasons why we should bathe.
12. When is it best not to bathe?
13. What do you think of cosmetics and hair-dyes?

CHAPTER XXV.

TEMPERATURE OF THE BODY.

THROUGHOUT the heat of summer and the cold of winter, the body maintains the same temperature. You may be in a very hot room one hour, and in a cold one the next; but the bodily heat will be at a fixed point; namely, 98½° F. This heat is the result of the many changes that are taking place in the tissues of the body.

Warm-blooded animals, as men, dogs, birds, etc., are those whose temperature is generally above that of the air surrounding them ; while cold-blooded animals have a temperature about the same as that of the air or water surrounding them.

To Cure a Cold. Usually after we have taken cold the skin is inactive. Hot flashes and chilly sensations creep over the body. A cold may frequently be broken up by restoring activity to the skin. Hot drinks, as hot lemonade or hot ginger tea, and hot foot-baths, are simple remedies that will excite the skin and send the blood to the surface again. Such simple treatment as here suggested may prevent a long and serious illness.

The Objects of Clothing. Food supplies heat to the body; and we have already said that animal heat is caused by the changes going on in the tissues, — by the changing of food into tissue. One object of clothing is to prevent too great a loss of heat from the body; for the air surrounding the body is nearly always cooler than the body itself.

Clothing and Food. It follows, then, that clothing answers a purpose similar to that of food. Food produces heat, and clothing prevents its escape. Therefore poorly fed persons need more clothing in winter than those who are well fed. Men and animals do not need so much food when kept warm as they do when chilly and cold, and they can do much better work if well protected from the cold. Clothing protects the body from external injurious substances, and from the storms of rain, wind, or snow. It is moreover an ornament to the body.

The Clothing of Animals. The lower animals have no choice as to what they shall wear. Nature gives them an abundant covering, which in some cases is very beautiful. She sometimes changes their clothing for them, as when the horse sheds his heavy coat of hair in the spring, that he may have a lighter one during the heat of summer.

What shall we wear? But man is left to make his own choice of clothing. What shall we wear? The answer to this question has been the occasion of no small amount of study and care. Is it warm, — what shall we wear? Is it cold, — what shall we wear? Are we going to a party, — what shall we wear? The decision that has been given has cost, in too many cases, most severe pain, and even death itself.

Clothing for the Head. The covering for the head is often either too heavy or too closely fitted. The scalp becomes heated. There is no proper opening for the hot air to escape, and headaches are of frequent occurrence. Sometimes the men are inclined to speak lightly of the small bonnets of the women; yet in many respects such coverings are far superior to those worn by their critics. The small bonnet does not confine the air over the head, it does not produce heat, and is not too closely fitted to the scalp.

The Weight of Clothing. Heavy clothing is not necessarily warm. The cloud worn by ladies as a covering for the head is very light, and yet it is very warm. There is a great deal of air confined in its meshes, and air is not a good conductor of heat.

The shoulders should, so far as is possible, be made to bear the weight of the clothing. Then the lungs and heart above are not compressed, and the stomach and liver below are not affected. Clothing worn in this manner does not prevent the free circulation of the blood, does not compress any organ, and does not interfere with the natural, graceful movements of the body.

Too high a Temperature. We should avoid overheating the body, and thus causing it to become weak; this is often done by wearing too much clothing, and by living in rooms that are kept at too high a temperature. The thermometer should not register over 70° Fahr. in our living rooms.

The Clothing should be changed. All the clothing should be changed at night. Whenever you have been caught in a storm, and your clothing has become damp or wet, change it as soon as possible for dry. In the mean time it is wise by some exercise, as quick walking, to keep the body from getting chilled. A brisk rubbing of the skin will aid in restoring it to its normal condition after the chilling effects of the dampness, and thus possibly prevent a cold.

THE EFFECTS OF ALCOHOL ON THE TEMPERATURE.

The question we have to answer here is this: Do alcoholic drinks raise the temperature? What truth is there in such a statement as this: "It is a cold day, we must take something to warm us;" or, "We must take some wine with us, as it will be a long drive, and we shall get very cold." These statements are made on the assumption that alcohol warms the body; in other words, that it raises the temperature of the body. Is there any truth in this supposition?

This necessitates a careful study of our subject.

The first effect of alcohol on the system, as we have already learned, is one of excitement. The capillary blood vessels become distended with blood, more blood goes to the capillaries of the stomach, more goes to the skin, and the surface of the body becomes distinctly warmer. This stage of excitement causes an increase in the temperature. This is true of all animals, including man. This period is a brief one, yet it has given to alcohol the credit of being able to warm the body. The undue amount of blood alcohol sends to the surface is cooled there, and on its return, the heat of the whole body is lowered.

Soon a great change takes place. The brief period of excitement is followed by a period of much longer duration, during which there is a rapid decline in the bodily heat. In birds this may eventually amount to as much as five degrees. In dogs the fall is not quite so much,— about three degrees. In man, the fall is often as great as two degrees, and in excessive drunkenness it may be much more. The normal standard may be reached again in a few hours if the amount of alcohol taken is small ; but if enough be taken to cause prolonged sleep, several days may pass before the normal degree is reached again.

The great lesson to be learned is this: that after a brief period of excitement there is always a reduction of the animal heat. Therefore alcohol will not aid us to withstand the cold ; on the contrary, it will cause us to suffer more from the effects of the cold.

Testimony of Travellers. The Arctic explorer Dr. McRae says: "The moment that a man had swallowed a drink of spirits it was certain his day's work was nearly done. In that terrific cold the use of liquor as a beverage when we had hard work on hand was out of the question."

Another Arctic explorer, Adam Ayles, says that those of his men who used liquor occasionally did not bear the work of that extremely cold country nearly as well as those who never took a drop.

Still another explorer, Dr. John Rae, says: "In nearly all the cases of death that I have heard of, it was found on inquiry that the persons so dying had taken some alcoholic drink."

As a conclusion of this whole discussion, let us notice what one of the greatest physiologists has recorded. In a work on The Physiology of Alcoholics, by Dr. W. B. Carpenter, of London, is the following passage:—

"My first illustration was from Sir John Richardson, a medical officer high in our naval service, who was early associated with Sir John Franklin in his Arctic explorations. It was, then, his conclusion that, even under extreme privation, the use of alcoholics did much more harm than good; so that it was better to burn the alcohol in a lamp, and to heat tea or some other liquid with it, and by drinking this to get some real heating effect, than to put the alcohol into the stomach. For what heat they got from one was so much gain; while the other, being only a stimulant, was followed by a depression which made the cold seem only the more severe.

"On another expedition Sir John Richardson passed the

winter with a party in the north of America, as near the border of the icy sea as they could reach. They were well supplied with food, and lived in a log house which had been built for them by our Hudson's Bay Company. Sir John had made it a strict condition that his party should go out upon strictly total abstinence principles; he would not have any spirits at all. It was a part of his work to make a series of magnetic observations, and the magnetic observatory was a short distance from the house. Sir John said he was accustomed to go out at night from the house to the observatory without even putting on his overcoat. I asked him how cold was the temperature to which he exposed himself. He said that the temperature in the log house was about fifty degrees above zero, and that outside it was fifty degrees below zero. Here was a change of one hundred degrees, which he found he was enabled to endure for one quarter of an hour without an overcoat."

QUESTIONS.

1. Tell about the temperature of the body.
2. Give some good remedies for a cold.
3. What is the object of clothing?
4. How may clothing affect the amount of food?
5. Do animals change their clothing?
6. When should we change our clothing?
7. How does alcohol affect the temperature of the skin?
8. Does this condition last long?
9. What follows this period of excitement?
10. What is the great lesson to be learned from this?

CHAPTER XXVI.

THE NERVOUS SYSTEM.

THE nervous system consists of the brain, the spinal cord, and the nerves.

The Brain. We call the brain the seat of intelligence. It is a jelly-like mass, surrounded by membranes, and well protected from injury by the bones of the skull. It is full of blood vessels; some of these are quite large, but the majority are minute capillaries. The surface of the brain is not smooth, but is thrown into ridges, between which are depressions.

The White and the Gray Matter. The brain consists of two kinds of matter, — the white and the gray matter. The former is composed largely of nerve fibres, while the latter consists principally of nerve cells. The gray matter is on the outside of the brain. It is the gray matter that commands, while the white matter obeys. The gray matter originates, and the white matter conveys the messages.

The Cerebrum. The part of the brain above the ears is called the cerebrum, or the great brain; the part at the back of the head, beneath the cerebrum, is called the cerebellum, or the lesser brain. It is supposed that the cerebrum is the organ of the mind. Here we think, and know, and reason.

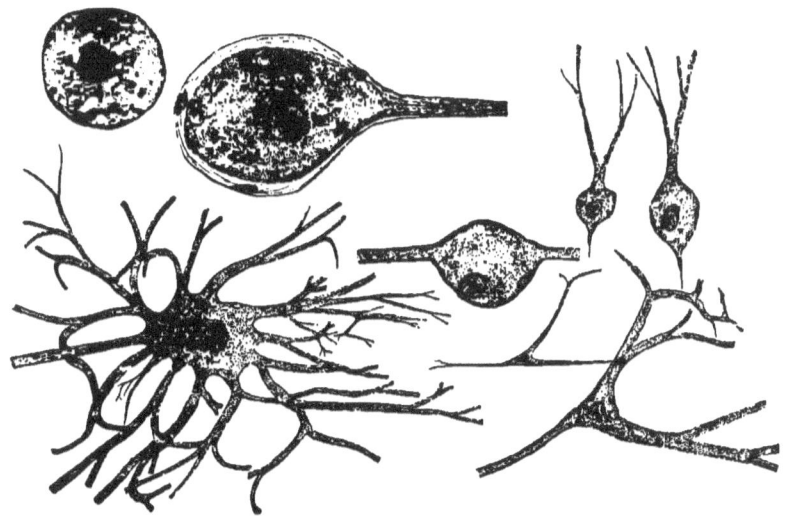

FIG. 53.   Various forms of nerve cells.

The cerebrum is divided into two parts by a natural fissure in the middle line, passing from the front backwards. From this it would seem that we have two brains, — a right brain and a left brain. But at the bottom of this fissure the halves are united by a band of nervous tissue (Fig. 55, 3); so doubtless their action is in some way connected. The ridges, or convolutions, of the cerebrum vary in different animals. As a rule, the more intelligent the animal, the more numerous are these convolutions; the deeper the depressions between them.

The Weight of the Brain. The brain of the elephant is the heaviest known, which weighs from eight to ten pounds. The brain of the whale comes next, which weighs from six to eight pounds. The average weight

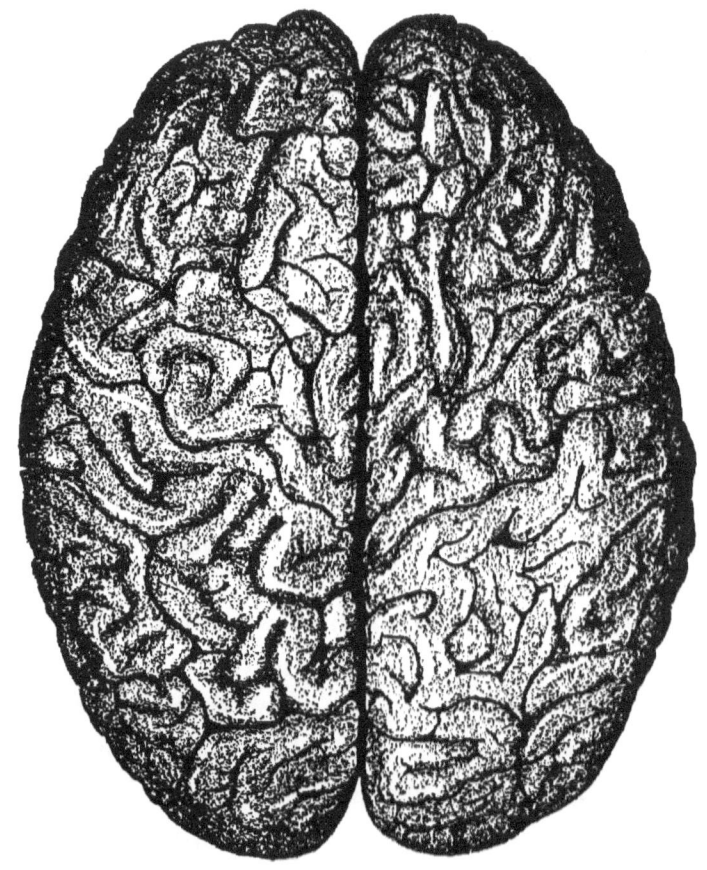

Fig. 54. The human brain, viewed from above : only the cerebrum is seen.

of the human brain is, for the male, a trifle over three pounds, and for the female, about one third of a pound less.

Something more than Brain needed. It seems sometimes that a man may have a large and fine brain, a healthy body, and be in every way fitted to succeed in life, and yet lack the desire to put his powers into action. We say such a person has no ambition. He is disposed to

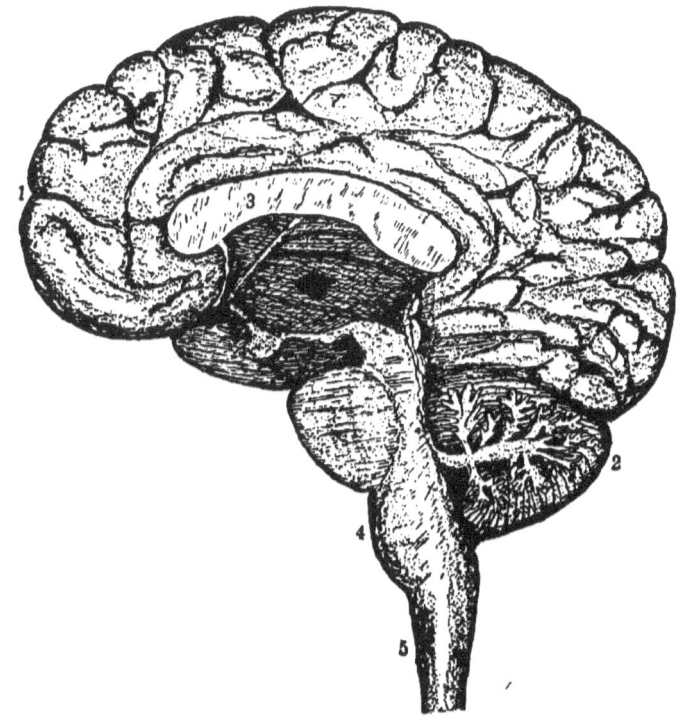

FIG. 55. One half of the brain, — the inner surface: (1) the cerebrum; (2) the cerebellum; (3) the band of tissue that unites the two sides of the brain; (4) the medulla; (5) the spinal cord.

take life too easily. We must not only have a healthy body and a fine brain, but we must also have a desire to work, — an ambition to be among the best in all we undertake.

The Cerebellum. The cerebellum is beneath the back part of the cerebrum. It has no convolutions, although there are ridges running over its surface parallel to each other.

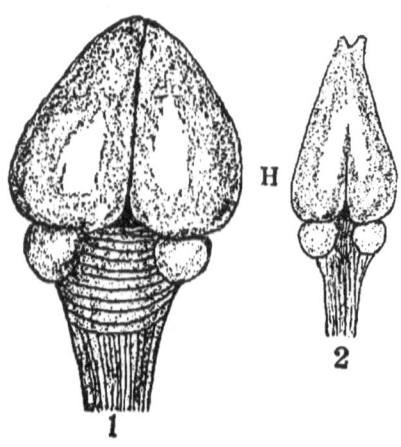

FIG. 56. (1) Brain of pigeon; (2) brain of frog, — both viewed from above. There are no convolutions on the cerebrum.

Have the Nerve Centres the Sense of Feeling? The cerebrum and the cerebellum may be cut, and portions of them removed, without causing pain. They have not the sense of feeling.

The Medulla Oblongata. At the upper end of the spinal cord, and between it

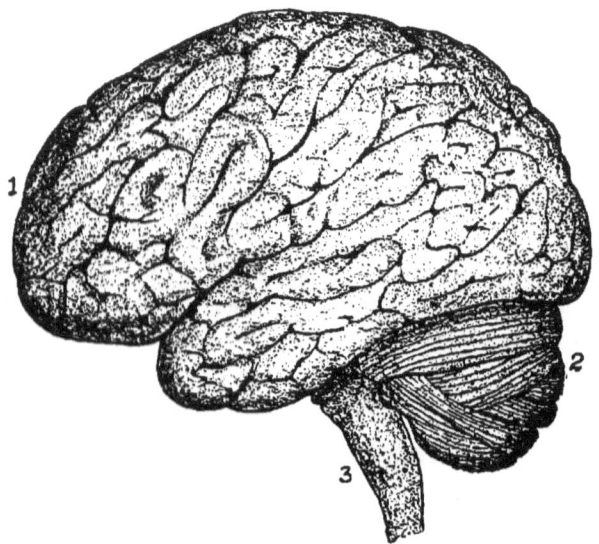

FIG. 57. Side view of the whole human brain : (1) cerebrum; (2) cerebellum; (3) medulla.

and the brain, is an enlarged continuation of the spinal cord, called the medulla oblongata, which is well protected by the thick bones at the base of the skull. This part of the brain, which controls the breathing and other important func-tions, must be regarded as the most delicate portion of the whole body. The cere-brum, or the cerebellum, may be entirely removed without destroying life for some time. Portions of the cerebrum, cerebellum, or spinal cord may be removed, and recovery follow. But if the medulla be destroyed, death will follow instantly. In fact the prick of a needle on a certain part of it is sufficient to cause death.

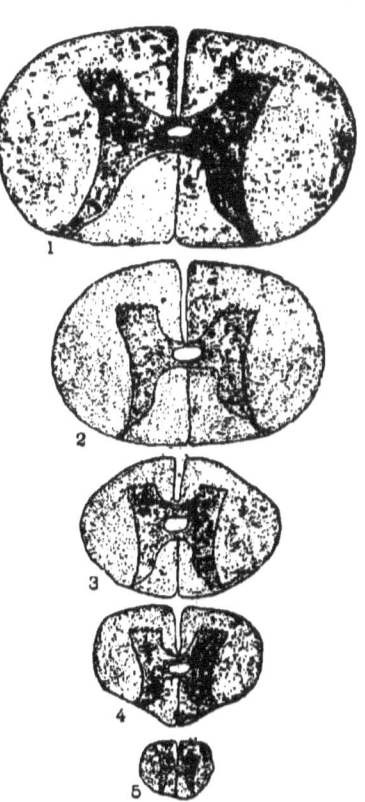

The Spinal Cord. The spi-nal cord is nearly circular in shape, about eighteen inches in length, and half an inch in thickness. It connects

FIG. 58. Cross sections of the spinal cords of different animals, repre-sented as twice the natural size: (1) horse; (2) ox; (3) man; (4) hog; (5) squirrel.

above with the medulla, and terminates at the lower end of the spinal column in a number of fine threads. (Fig. 59.) The spinal cord, like the brain, is divided into halves. A fissure extends down its front and another down its back, nearly dividing it. (Fig. 60.) From each half, or each side, of the cord there are given off thirty-one small cords, or nerves.

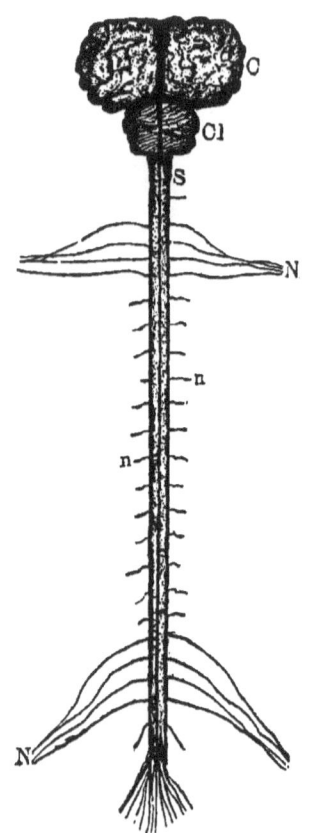

Structure of the Cord. The cord is composed of gray and white matter, similar to that found in the brain. The gray matter is in the centre of the cord, and is collected together in such a manner that when the cord is cut across the gray matter resembles the letter H. The nerve fibres go up and down the cord, and carry messages to and from the brain. After entering the brain these fibres are distributed through all parts of it, and are connected with the nerve cells.

The Nerve Centres. The brain and the spinal cord are called the nerve centres. From these the majority of the nerves take their origin.

FIG. 59. View of the brain and spinal cord: c, cerebrum; cl, cerebellum; s, spinal column; N, nerves for the arms and legs; n, nerves going to the muscles and skin.

The Nerves. There are two kinds of nerves, — the sensory, or fibres of feeling, and the motory, or fibres of motion. The sensory nerves carry messages from different parts of the body to the nerve-centres. They convey to the brain such impressions as are produced on the tongue by anything sweet or sour; or on the ear by harsh or gentle sounds; or on the eye by colors. These impressions are made on the ends of the nerves; they are

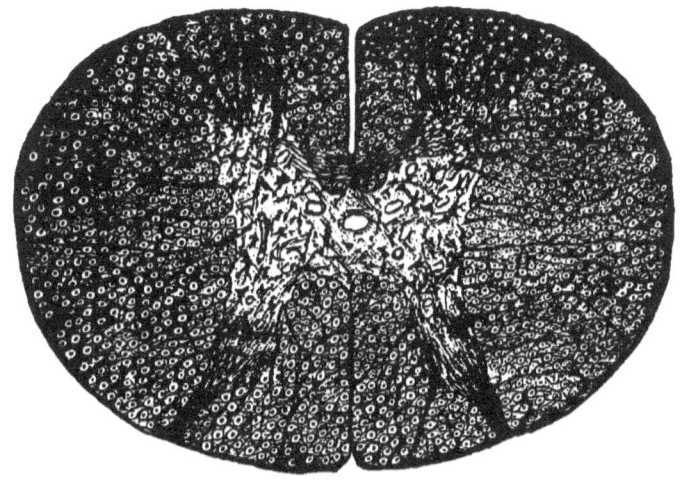

FIG. 60. A cross-section of the spinal cord, as seen with a microscope.

then carried by the nerves to the spinal cord; they are then conveyed up the cord to the brain, where, in some mysterious way, we are informed of the facts going on at the ends of the distant nerves.

The motory nerves convey messages from the nerve centres to the muscles, causing them to contract.

Each one of the spinal nerves is connected with the cord by two roots. One root enters the front of the cord, and is composed of motory nerves; the other enters the back of the cord, and is composed of sensory nerves. (Fig. 61.)

Reflex Action. Reflex action is well illustrated by irritating the foot of a person who is asleep. The foot will

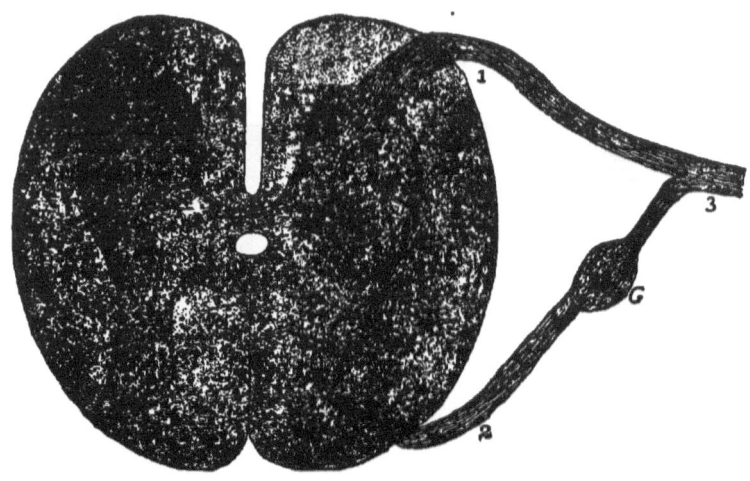

FIG. 61. A diagram illustrating the origin of the spinal nerves from the spinal cord ; (3) is a spinal nerve; (1) and (2) are the roots, which originate from the gray matter of the cord ; G, a collection of nerve cells: (1) the motory root, (2) the sensory root.

be quickly drawn away from the threatened injury, and yet the mind will know nothing of it. Fig. 62 illustrates

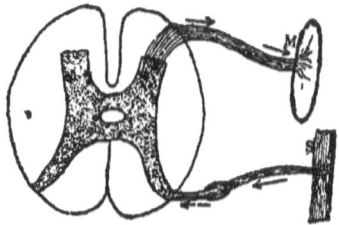

FIG. 62. A diagram illustrating reflex action : s, the skin ; M, a muscle. The arrows represent the direction of the nerve-current during a reflex act.

how the impression goes from the skin of the foot to the spinal cord, and then down to the muscles.

The spinal cord is one great reflex centre. Impressions are carried to it by the sensory nerves, and, without going to the brain, they are at once sent out to some muscle through the motory nerves. Even the will is not strong enough to control all the reflex acts, for if we inhale pepper, we either cough or sneeze, and cannot prevent it. It is purely a reflex act. We may determine not to sneeze, but it is of no use.

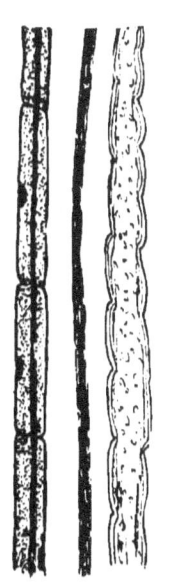

Fig. 63. Nerve fibres, magnified.

The Nerve Current. The peculiar power carried by the nerves is called the nerve-current, or nerve force. It travels along the nerves at the rate of over one hundred feet a second. We do not know what this nerve force is, although we can interfere with its action. We may be seated in a cramped position, so that some nerve is pressed; the nerve current is disturbed, and when we try to move the limb it gives pain. We say the limb is asleep. It may be impossible to move the limb at first, owing to a complete stopping of the current.

Study is a Healthful Exercise. The cerebrum is generally regarded as the seat of the mind. We know that there is a very general law that the proper exercise of a part tends to make it grow and develop. It is possible, therefore, by properly exercising the brain, to aid the growth and development of the intellect. While the mind can be greatly improved by exercise, it can also be injured either by over-exercise or idleness.

QUESTIONS.

1. Of what does the nervous system consist?
2. Describe the brain.
3. What composes the white matter? The gray matter?
4. Which is on the outside of the brain?
5. Which commands and originates? Which conveys messages?
6. Where is the cerebrum? The cerebellum?
7. Which is the organ of the mind?
8. How is the cerebrum divided?
9. What is said about the convolutions?
10. How heavy is the human brain? Is this the largest?
11. What is needed besides brains?
12. Have the nerve centres any feeling?
13. Why is the medulla so important?
14. Describe the spinal cord.
15. How many kinds of nerves are there?
16. What is the use of each?
17. Illustrate reflex action.
18. Tell about the nerve current.

CHAPTER XXVII.

ALCOHOL, TOBACCO, OPIUM, AND THE NERVOUS SYSTEM.

IF the effects of alcohol on the nervous system were more fully understood, its use would be almost entirely abandoned by physicians, and it would be generally discarded as a beverage.

There are many persons who begin to take some form of alcoholic drink because they think it may do them good. They are in poor health, and so they take this as a remedy, or a tonic. Soon, however, the habit is formed, and then they are certain they cannot live without it.

One of the most striking things about alcohol is that it has a special affinity for nervous tissue. We mean by this, that after an animal has died from the effects of alcohol, and its different tissues have been examined by the chemist, it is found that there is more alcohol in the nerve centres and the nerves than in any of the other tissues. This is a forcible statement, and it comes from the highest medical authorities.

After the death of a person who has used alcohol largely, alcohol has been found in the fluid that is normally in the brain; and it has actually been distilled from the brain.

When alcohol is taken into the system it is carried to all the tissues, but *more goes to the nervous tissue than to any other.* Its effect on nervous tissue is just what might be expected when we remember that alcohol is a narcotic, — it paralyzes it.

If alcohol really paralyzes nerve tissue, why does it appear at first to excite the person who takes it? Every increase in the supply of blood to the brain or other organ is associated with an increase in the activity of that organ. Alcohol partly paralyzes the nerves that should control the movements of the blood in its vessels, and these vessels stretch, letting too much blood flow to the brain, which is thereby excited. But this excitement is not the healthy, trustworthy activity of a sound brain, but an untrustworthy mental condition, in which the judgment, the reason, and the conscience do not control the words and the conduct as they should.

The relaxing of the blood vessels through the influence of alcohol causes the heart to beat faster. As the first effect of alcohol is thus to increase the action of the heart, and as this causes an extra flow of blood to the face, so it leads to similar action in the brain. There is too much blood present, and for a short time this excites the nerve cells to increased action.

Dr. Richardson, of London, once saw a man, while under the first exciting effects of alcohol, beheaded by a railway car. The bones of the skull were so broken that the complete brain lay open to view. It was not three minutes after death before the doctor was making a careful examination of it. He says that the odor of alcohol was most distinct. The membranes around the brain were as red as if they had been injected with vermilion. The white matter was so charged with blood that it could scarcely be recognized as white. One of the membranes surrounding the brain was so filled with blood that it did not look like a membrane at all, but rather like a thin sheet of clotted blood.

After the period of excitement has passed, another effect appears. The person has to think of his walking, as he cannot think of something else and walk also. He has to think intently about the movements of his limbs, in order to make them correct. This shows that the spinal cord is coming under the influence of the alcohol. It is during this stage that the muscles of the lower lip begin to fail, and the muscles of the lower limbs become weak. Severe vomiting is likely to occur at this time, especially in young persons. It is an effort of nature to throw off the harmful invader, and is usually followed by relief.

With the increased effects the cerebrum acts in a very disorderly way. The sentences are broken, and the ideas confused; the nerve cells are over excited, and the influence of the will is less and less powerful. Reason has stepped aside, and now the merely animal part of the man's nature assumes control. The cruel man becomes more cruel; the untruthful, more false; and the vile man shows his true disposition. The muscles respond to stern commands but feebly, and the limbs scarcely support the body. Words cannot be formed, and no articulate sounds escape. If these effects extend but a little farther, the brain is completely overpowered, the voluntary muscles cease to act, sensation is lost, and the body becomes a mere senseless mass of flesh.

Has the body any life? Yes. The heart still beats, and the chest still heaves. Enough blood is sent to the lungs to get the oxygen required to keep the body alive.

As a rule the brain loses its power before the heart; that is, when death occurs from almost any cause, the person will become unconscious before the heart ceases to beat. So alcohol will make the person entirely unconscious before it will stop the beating of the heart; thus the life is spared. He cannot drink more until the body recovers sufficiently to restore consciousness. Did the heart fail before the brain, the number of immediate deaths alcohol causes would be increased a hundredfold.

" But," it is said, " few go so far as to become intoxicated. The question is simply this : What effect do moderate and frequently repeated doses of alcohol have on the nervous system?" We answer as follows : —

There is without doubt a direct effect of alcohol on the nerve tissue itself. As alcohol is found in the brain, so it must affect the cerebral matter by direct contact with it. Alcohol also affects the brain by inducing changes in the supply of blood.

Dr. Bartholow, a well-known medical writer, says in his work on " Materia Medica," in speaking of the effects of alcohol as a medicine, in the case of the patient who has long been subjected to its influence, that

" the nerve cells of the gray matter are more or less fatty and shrunken." He further says that as a result of the shrinkage of these cells, and of the other tissues also, "the whole cerebrum becomes smaller, and the space thus made becomes filled with a watery fluid."

Shakspeare, in the tragedy of " Othello," says, " Oh, that men should put an enemy in their mouths to steal away their brains ! " and temperance writers have frequently quoted him ; but they little thought that one of the most noted medical men of this century would describe, in a work written solely for medical men, in a cool and logical way, the effects of this drug on the system, and would in that description confirm their statements. Temperance teachings and science are surely in accord. " Alcohol shrinks the brain," says Dr. Bartholow ; and he further states that its use results in " impaired mental power," " muscular trembling," and a " shambling gait."

Dr. William A. Hammond, the eminent New York specialist, and one of the widest known authorities on nervous diseases, has made experiments to ascertain the effects of alcohol on the nervous system. If there were good in it, he wished to know what it was, and let his patients profit by it.

In 1887 he wrote to the Hon. H. W. Blair as follows : " Weighing all the points, for and against, mankind would be better mentally, morally, and physically if the use of alcohol were altogether abolished."

DELIRIUM TREMENS.

This disease may be caused by a single intoxication, but usually it is the result of the excessive and long-continued use of alcoholic drinks. It may attack a person of very nervous temperament, when comparatively a small quantity has been taken. It is one of the most terrible of all the effects of alcohol. The victim is wild with fear and dread. He sees multitudes of horrible creatures all about him to do him harm. The faces of his dearest friends now become like so many fearful, ugly monsters; he strikes himself and his friends; he is choking from thirst, and is constantly calling for more drink. There is no way of telling whom, among those who use alcoholic drinks, it will affect, nor when it will come. The attack may be postponed until after years of continued drinking, or it may attack the one who has but recently become a victim to the appetite.

THE EFFECT ON THE MIND.

One of the first, and yet one of the most terrible, effects of alcohol is that it greatly weakens the will-power. The slightly intoxicated man goes about the streets singing and laughing in a very silly way. He is easily provoked, does very rash things, and becomes an object of sport to thoughtless boys.

Consider a man that is honest and kind when sober.

Who can tell what he will do when under the influence of the alcohol poison? If the drinking be continued, the will power must sooner or later suffer. The passion for drink becomes greater and greater, until nothing will be thought too dear to exchange for it. The records of the various crimes show that a large percentage of them were committed while under the influence of liquor.

The United States Commissioner of Education says: " From eighty to ninety of every one hundred criminals connect their crimes with intemperance. One of the judges of the city of Philadelphia says that four fifths of all the crimes committed in that city are due to the influence of strong drink." We could give a great many pages of such testimony, showing that the effect of alcohol is to lead men to commit deeds which they would shrink from if they were in their right mind.

While such results follow the free use of alcohol, the frequent use of small doses, not sufficient to produce intoxication, also affects the mind, weakens the sense of right and wrong, and therefore tends to destroy character itself.

TOBACCO.

When once begun, the use of tobacco is quite likely to be continued, so strong is the habit. It frequently produces a feeling of nervous restlessness, which may lead to the use of alcohol. It is well-known that one narcotic habit leads to another. Thus the narcotic tobacco often creates an appetite for the more dangerous narcotic alcohol.

OPIUM.

Opium is more pronounced in its effects than tobacco, and differs both from it and alcohol in this marked respect,—it may be used for years, and the secret be known only to the user. The odor of tobacco betrays its victim even when afar off; the teeth and the corners of the mouth betray the chewer, while the smoker is enveloped in a cloud of his own creation. The victim of alcohol may be known by the unsteady gait, the bleared eye, the thick tongue, and the terrible breath. But the victim of opium has the secret all to himself. The opium habit has not become general in this country, and it is to be hoped it never will; as its effects are to be dreaded even as much as those of alcohol. The power of opium to relieve pain is wonderful, hence there is great danger that the opium habit will be formed whenever a person has long-continued suffering. Opium affects all parts of the body, but especially the nervous system. Under its continued use the memory fails, and there is a partial paralysis of the lower limbs, giving a stooping or creeping appearance. A distaste for food follows, the stomach refuses to act, and in many cases death ensues.

The peculiar and most marked feature about the opium habit is that its victim has an almost irresistible desire to repeat the dose. The amount must be gradually increased to produce the desired effect, until sometimes the quantity taken is enormous. Under its influence the moral sense is destroyed, the mind weakened, and the whole body thrown into a dazed stupor, unfitting the victim for any sphere in life.

QUESTIONS.

1. Describe one of the most striking things about alcohol.
2. How is this known?
3. How does alcohol affect nerve tissue?
4. How does this paralyzing effect appear to excite the person? Explain this.
5. Why then is it an error to suppose that alcohol will help brain work?
6. How does the effect of alcohol on the heart affect the brain?
7. How may we know when alcohol is affecting the spinal cord?
8. How does the cerebrum finally act?
9. How is this shown?
10. What would happen if the heart lost its power before the brain from the effects of alcohol?
11. Why must alcohol have a direct effect on brain tissue?
12. In what other way does alcohol affect the brain?
13. What is Dr. Bartholow's testimony?
14. What does Dr. Hammond say concerning the use of alcohol?
15. How may delirium tremens be caused?
16. Describe some of its symptoms.
17. How does strong drink affect the will power?
18. What shows that drinking leads to crime?
19. What does the Commissioner of Education say?
20. What testimony does a judge give?
21. What results follow the frequent use of small doses?
22. To what does the tobacco habit often lead?
23. What are the effects of opium on the nervous system?

CHAPTER XXVIII.

THE SENSE OF SIGHT.

The Eye is well protected. The eye is well protected in its deep socket of bone. The brows project over it, and are covered with thick hair, which lies in such a direction that the perspiration from the forehead will be carried to one side of the face, and will not run into the eyes. The nose serves to protect the eyes from injury. Directly in front of the eyes are two curtains, which can be quickly and freely moved. These are the upper and lower lids. The upper lid is much more movable than the lower. The lids have a row of delicate hairs on their edges, called eyelashes. These are of great use to the eye, even in the dark; for if an insect or any particle of matter comes in contact with them, the eyelids close at once, preventing the harmful object from touching the eye itself.

The eyelids protect the delicate eye from heat and cold, they keep out dust and dirt, they regulate the amount of light to be admitted to the eye, and they keep the front of the eyeball moist.

The Oil Glands. There are glands in the eyelids which secrete an oily substance that flows over the edges of the lids and keeps them from adhering to one another. It also tends to keep the tears from running over the edges of the lids and down upon the face.

FIG. 64. The eyelids of the right eye, viewed from the inside: (1) the lachrymal gland; (2) the oil glands in the eyelids.

The Tears. There is a small gland in the outer and upper part of each orbit. This is called the lachrymal gland, and it secretes the tears. The secretion is a constant one, and the fluid is distributed over the front of the eyeball by the movements of the eyelids. The tears are carried away from the eyes through two openings. There is one opening on each lid near its inner extremity. The one on the lower lid may easily be seen by looking in a mirror and slightly pulling the lower lid down. It has the appearance of a very small dark

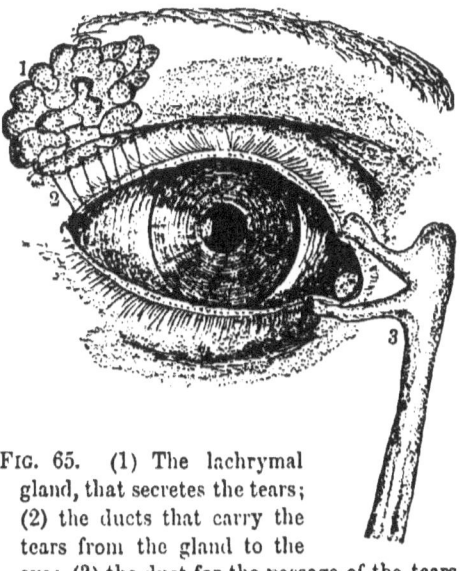

FIG. 65. (1) The lachrymal gland, that secretes the tears; (2) the ducts that carry the tears from the gland to the eye; (3) the duct for the passage of the tears to the nose.

spot. (Fig. 65.) The two openings of each eye enter a duct leading to the nose, into which it opens. Generally the tears pass through this duct, but sometimes more are secreted than can pass through it; then they flow over the eyelids and down upon the cheeks.

The Eyeball. The eyeball is a round body with many membranes, or coats, surrounding it. The figure shows that it is not perfectly round, for the front part protrudes more than the other parts. This front and highly curved part is transparent, and through it the light readily passes.

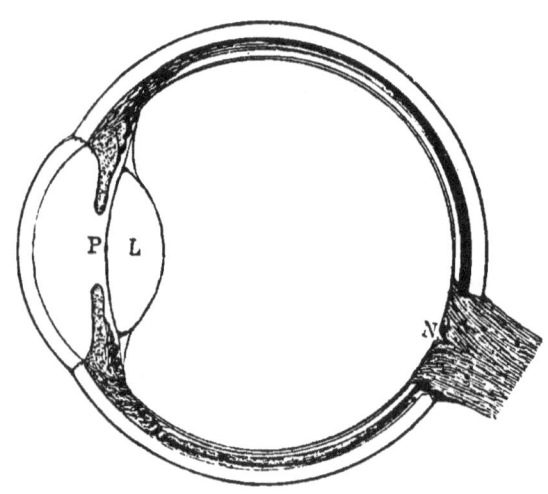

FIG. 66. P, the pupil; L, the lens; N, the optic nerve.

The colored circle of the eye, which makes it appear black or brown or blue, is called the iris. It is found in the middle of the eye, with a circular hole in its centre, called the pupil.

By the action of certain muscles the size of the pupil may be changed. If the light is too bright, the pupil will be made smaller, so that but little light will reach the back of the eye; but if the light is very dim, then the pupil will enlarge, in order that as much light as possible will reach it.

The pupil is not found in all animals; in the cat it is simply a vertical slit when it is contracted, but round when it is fully dilated. It may become so large in this animal that enough light enters the eye to enable it to see even when we think it is dark.

How we see. The light enters the front of the eye, passes through the pupil, then through the lens, and lastly strikes at the back of the eye a delicate membrane which is directly connected with the brain by means of the optic nerve. When we look at an object, an image of it is made on this delicate membrane, and our brain sees the image. If we cut the optic nerve, and thus sever the connection between the eye and the brain, the image is made on the eye exactly as before, only the brain does not perceive it. So the eye simply acts like the camera of the photographer. The photographer sees, not his camera. It is the brain which sees, after all.

Care of the Eyes. There are many causes of trouble with the eyes. If the whole body is weak, the eyes also are likely to be weak. They are easily tired, and the ordinary use of them causes them to become inflamed. As inflammation of the throat arises from a cold, so the delicate membrane covering the front of the eye becomes inflamed from the same cause. The eyes feel as if there were sand in them, and at times they are used only with great pain. Whenever the eyes are inflamed, or their use causes pain, or when reading causes headache, a competent physician should be consulted, to ascertain what the trouble is, and to remedy it, if possible.

It is a terrible misfortune to be blind, and very sad to have the eyes so weak that the pleasure of reading and studying must be denied. We should therefore use every possible care to prevent any such misfortune overtaking us. The following suggestions may aid one in preserving good eyesight : —

Do not read by twilight.

Do not use the eyes when they feel tired.

The light should be clear and steady.

It should be neither a dim nor a very strong light.

Never look directly at a brilliant light, as the sun.

Do not read when lying down; the upright position is the natural and proper one.

Do not read when riding in the cars or in a carriage, or while walking. The eyes become quickly tired from the irregular muscular action.

Do not look too long at any one thing. Rest the eyes by looking around at frequent intervals.

Squinting may result in serious trouble, as it strains some of the muscles of the eye.

Do not face the light when reading in the evening. Put the shade on the lamp, and let the light come from over the shoulder.

Do not allow strong sunlight to fall on the eyes upon first awaking in the morning.

A book should be from twelve to sixteen inches from the eye when reading.

Have any object that may have fallen into the eye removed as quickly as possible.

Never rub the eyes to remove dust or dirt. These may be removed by carefully wiping with a folded corner of a soft handkerchief.

Avoid reading and studying if the eyes are inflamed.

Never use ointments or eye-washes without the advice of your physician.

The eye is a delicate organ, and we cannot fully estimate its value to us. It is therefore extremely unwise to neglect it in any way, or to attempt to cure it of any defect, without the advice of a physician. We should use every possible means to keep it in perfect health, and when it shows signs of weakness we should at once seek proper medical aid.

ALCOHOL AND THE EYES.

The bleared eyes of the hard drinker are one of the tell-tales of his habits. The blood vessels are filled to their utmost with blood, making the eyes look blood-shot. This inflamed condition is likely to continue as long as alcohol is used. After the habit is broken, and no more strong drink is taken, under proper treatment the inflammation may be relieved or entirely cured.

The enlarged vessels noticed here are doubtless due to the same cause as the enlarged vessels of the nose; namely, a paralysis of the nerves controlling the blood-vessels.

Persons who indulge in the use of strong drinks are very likely to have weak and inflamed eyes.

TOBACCO AND THE EYES.

The smoke of burning tobacco is very irritating to the eyes, and red edges to the eyelids are very common among smokers. In serious cases arising from the use of tobacco there are sharp pains in the eyeballs, and vision is greatly affected. There is no help for these troubles until the person stops the use of their cause.

QUESTIONS.

1. Describe how the eye is protected.
2. Of what use are the lids?
3. Describe the oil glands.
4. Where are the tears secreted?
5. How are the tears carried away from the eyes?
6. Describe the eyeball.
7. What is the iris?
8. What is the circular hole in its centre called?
9. How does light affect the pupil?
10. Where is the image of an object made?
11. What is the effect of cutting the optic nerve?
12. To what can we compare the eye?
13. Can you give a few suggestions for preserving the eyesight?
14. How does alcohol sometimes affect the eyes?
15. Is tobacco ever injurious to them?

CHAPTER XXIX.

THE SENSE OF TASTE.

THE sense of taste is made possible to us largely by means of the tongue. The lining membranes of other portions of the mouth may have something to do with it, but it is principally due to the membrane on the upper surface of the tongue.

The Tongue. The tongue, which is composed of voluntary muscle, is easily moved in any direction. When the body is in a healthy condition, the tongue is moist and of a light red color. A dry tongue denotes fever, while a furred tongue is pretty sure evidence of some disturbance of the digestive organs ; a bright red tongue also is a symptom of disease. Therefore by the appearance of the tongue the physician can learn the condition of his patient.

The healthy tongue is covered, over its upper surface, with minute elevations, called papillæ. The largest of these are found on the back part of the tongue. They are arranged like the letter V, with the point of the V towards the back. Other and smaller papillæ are easily seen, scattered over the tongue. Some of these papillæ act as organs of touch, for the tongue has the sense of touch just as the skin. Other papillæ act as taste-organs, and are connected with the brain by means of a nerve.

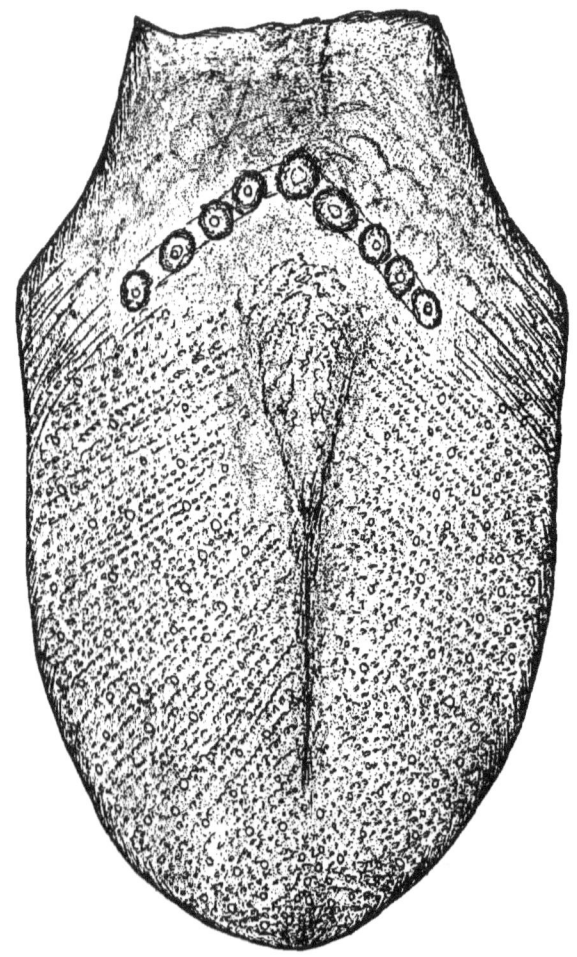

FIG. 67.   The tongue, showing the varieties of papillæ.

We cannot taste any substance until it is dissolved. If dry sugar is placed on a dry tongue, there will be no taste to it whatever. The saliva aids in dissolving the substance we taste, and the movements of the tongue bring the substance in contact with its papillæ.

Smell and Taste. The sense of smell frequently confuses the sense of taste. We think we taste a thing when really it is the odor which is the more prominent. We dislike to take medicine because of its unpleasant taste; so we follow custom and close the nose, and it is taken without trouble. In this case it is the odor we wish to get rid of rather than the taste.

Flavors. If the eyes and the nose are closed, the taste of an onion may be mistaken for that of some pleasant fruit.

We may become so accustomed to the taste of certain articles of food as to think them very agreeable, although at first they might be very disagreeable to us. Many persons at first dislike the taste of tomatoes or olives, or even oysters, but soon they acquire a great liking for them.

Only one flavor can be appreciated at a time. If more than one be tried at a time, the result is a confused taste. A strong flavor may so affect the parts that a weaker flavor may immediately follow it without being noticed. Advantage is often taken of this in giving medicine. A strongly flavored fluid is taken into the mouth, and immediately after the disagreeable dose is swallowed, followed by more of the strong flavor first used.

The practice of eating cloves and other spices is a very harmful one, as it is likely to destroy the sense of taste and seriously disturb the action of the stomach.

TOBACCO AND TASTE.

Tobacco has the power to blunt greatly the sense of taste. The papillæ of the tongue may become so penetrated with it that there is a flavor of tobacco in the mouth continually; hence more flavors and stronger ones must be used, that they may be noticed.

QUESTIONS.

1. Where is the sense of taste located ?
2. Describe the tongue.
3. Give its natural condition ; its condition in disease.
4. What is found on the upper surface of the tongue?
5. Where are the largest papillæ ?
6. How do these papillæ act ?
7. Can we taste perfectly dry substances ?
8. Of what use to taste is the saliva ?
9. How are smell and taste confused ? Illustrate.
10. What is said about acquiring a taste for certain foods ?
11. What is said about strong flavors and weak flavors ?
12. What objection to the practice of eating spices ?
13. What effect has tobacco on the sense of taste ?

CHAPTER XXX.

THE SENSES OF SMELL AND TOUCH.

The Sense of Smell. When certain substances are brought near the nose they produce a peculiar sensation which we call an odor. It is difficult to describe the odor of objects, since we cannot see or measure it.

The sense of smell is of great service to us in many ways. When we are conscious of any bad smell in the air, we know that it is unfit to breathe ; when food has a tainted odor, it should not be eaten. Many of the disagreeable odors come from substances that are harmful to the body, while agreeable odors are frequently not only pleasant, but healthful also.

Some odors may be carried for a long distance. It is said that persons on board vessels at sea have detected the odor of cinnamon which grew on a shore over two hundred miles away.

The Sense of Touch. The sense of touch resides in little bodies found in the skin. They are seen in Fig. 48, at 4. These little bodies are attached to nerves which convey the sensation to the brain.

Acuteness of Sensation. We may easily ascertain for ourselves which parts of the body are the most sensitive. Take two pins and hold their points at least an inch apart. Press them lightly against the skin on the back of the wrist of another person, being careful that both points come in contact with the skin at the same time. Repeat the experiment, bringing the pins a trifle nearer together each time. Soon the sensation will be as if only one point was touching the skin. Make a note now how near together the points are. Try the same experiment on the inside of the first finger. It will be found that the points may be brought much nearer together before the two points will feel as one, — which shows that the sense of touch is much more delicate at this place.

The Education of Touch. The sense of touch may be highly developed. Blind persons can read their Bibles very rapidly by passing the fingers over the pages of raised letters; their sense of touch is so delicate that the form of the slightly raised letter is made very clear to them.

Feelers. Many of the lower animals are provided with " feelers," in order that they may detect an object near them. The whiskers of the cat are probably for this purpose. The skin of that animal is so covered with hair that it can be of but little use as an organ of touch; therefore Nature has given it special prolongations, called " feelers." These feelers are fastened deep in the skin, and are attached to nerves which convey impressions directly to the brain.

It is said that a cat cannot get out of a totally dark room if its whiskers have been cut off. It will run against the wall or articles of furniture, and will not make its way out by any opening left for it, while if its whiskers are uncut it will readily find its way out.

QUESTIONS.

1. Where is the sense of smell located?
2. How is this sense of service to us?
3. What can you say of air or food which has a bad odor?
4. What is said of the distance to which odors may be carried?
5. Where does the sense of touch reside?
6. How is the sensation conveyed to the brain?
7. How can the sensitiveness of parts be ascertained?
8. Is the delicacy of touch the same in all parts?
9. Can this sense be educated? Illustrate.
10. How is this sense provided for in lower animals?
11. What is said about the whiskers of a cat?

CHAPTER XXXI.

THE SENSES OF TEMPERATURE, WEIGHT, AND HEARING.

The Sense of Temperature. Whenever our bodies come in contact with a substance, we are able to tell whether it is hot or cold. The faculty by which we are enabled to do this is called the " temperature sense." This is situated in the skin, the mouth, the throat, and at the entrance of the nose.

The Sense of Weight. This is also called the "muscular sense." By it we are enabled to judge the weight of bodies. By long practice persons are able to detect very slight differences between bodies even so light as coins.

The Sense of Hearing. The organs of hearing are among the most difficult parts of the body to understand. We see only the outer ear, but the anatomist speaks of two others, — the middle and the inner ears.

The Outer Ear. The outer ear is the peculiarly shaped piece of cartilage on the side of the head by means of which we catch sound waves. This is very easily seen to be true of the lower animals: they always turn their ears in the direction from which the sounds come. Even man is governed by the same principle, and when his hearing is not acute he will place his hand behind his ear and push it forward, at the same time making it larger by adding to it the width of his hand. The

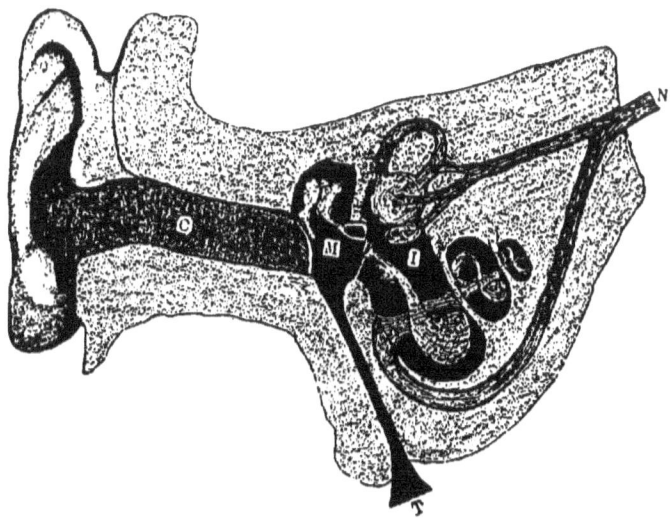

Fig. 68. The ear : c, the auditory canal, that leads to the middle ear; M, the middle ear, or drum. The drum-head is the curved white line to the left of the letter M; I, the inner ear; N, the nerve of hearing, going to the brain; T, the tube leading from the middle ear to the upper part of the pharynx.

auditory canal, which is a part of the outer ear, and leads to the middle ear, is about an inch in length.

The Middle Ear. At the inner end of this canal is the drum-head, while on the inner side of this membrane is the drum, or middle ear.

The drum-head completely shuts off communication between the outer and middle ears. When sound waves strike the drum-head, they cause it to vibrate, — much as the head of an ordinary drum vibrates when struck.

The middle ear, which is filled with air, is connected with the throat by means of a canal called the Eustachian tube. Through this tube air is admitted into the drum.

The Inner Ear. The inner ear contains the ends of the nerve of hearing, or auditory nerve. This part of the ear is carefully protected by the solid bones of the skull.

Causes of Deafness. We have spoken of a passage leading from the middle ear to the throat. We can understand now how it is that throat diseases often cause deafness, for the inflammation may readily extend from the throat along the Eustachian tube until it reaches the middle ear and affects it.

Blows on the ears are always dangerous, and may cause sudden and permanent deafness. Though the blow itself may not be severe enough to cause pain, yet the force of the compressed air against the delicate structures of the ear may occasion mischief. Another cause of deafness is the frequent cleaning of the ears with a pin or other hard instrument. This is likely to give rise to a slow inflammation, which may not show itself for months, or even years, but then gradual failing of the hearing will be noticed. Loud sounds, as the firing of cannon, also have caused deafness.

How to preserve the Hearing. Never use a pin or other hard substance to clean the ear. The ordinary washing and wiping with a towel are sufficient to insure perfect cleanliness. The wax that is in the deeper parts of the outer ear is useful for keeping out insects and dust. Do not allow cold air to blow into the ear; it is a frequent cause of catching cold. A good practice is to bathe the outer ear frequently with cold water. Inflammation of the ear is often occasioned by diving; it is well before plunging into water to place pieces of cotton in the ears. Never speak loudly in any one's ear. Should an insect get into the ear, have a competent person pour a little sweet oil into it; this will either kill the insect or drive it out.

QUESTIONS.

1. What is the sense of temperature?
2. Where is it situated?
3. What is the sense of weight?
4. Where is the sense of hearing located?
5. How much of the organ of hearing can we see?
6. What is the use of the outer ear?
7. How do the lower animals show this?
8. How can we catch more sound waves?
9. Where is the auditory canal?
10. What is at the inner end of this canal?
11. What is on the inner side of the drum-head?
12. What divides the outer from the middle ear?
13. How do sound waves affect the drum-head?
14. How is the middle ear connected with the throat?
15. What is this tube for?
16. What does the inner ear contain?
17. Give some of the causes of deafness?
18. Give a few directions how to preserve the hearing.

CHAPTER XXXII.

OPIUM, TEA, COFFEE, AND TOBACCO.

OPIUM is made from the juice of a plant called the white poppy, which is largely grown in India and China. Incisions are made in the unripe seed vessels, from which a white, milky juice escapes. This is allowed to dry, when it is gathered as a brown, thick gum. This gum is opium. From opium are made such drugs as morphine, laudanum, and paregoric.

Used very generally in some Countries. In China the people use opium as generally as Americans use tobacco. But in this country, as a rule, it is first taken, by the advice of the physician, to relieve pain.

How the Habit is formed. If the pain recurs as soon as the effects of the drug have passed off, another dose is taken, so that if the pain continues for weeks, and the use of the opium is continued also, the habit is formed, and the drug is taken long after all necessity for it has passed away.

Its Effects. Opium does not make its victim wild and uncontrollable, as does alcohol ; therefore it does not

show its effects to others as plainly as does alcohol.
But on the victim himself they are probably greater
than those produced by alcohol. Men become stupid
under its use, and gradually pass into a deep, heavy
sleep.

This deadly drug takes away the appetite, destroys
digestion, interferes with the action of all the organs
of the body, and powerfully affects the will. The opium
eater will sacrifice anything to get his favorite drug.
He loses all sense of honor and truth, and nothing is
too valuable for him to part with, if only it will procure
him his favorite drug.

None Escape. No one who uses opium escapes its bad
effects. It appears to affect all alike, weakening both
body and mind.

The Physician can detect the Opium habit. Its effects
on the body — as shown by the listless eye, the stooping figure, and the colored skin — enable the trained
eye of the physician to detect the opium eater with
almost unfailing accuracy.

Its Terrible Power. The power of opium to injure the
body is terrible enough; but when we consider that it
will weaken and break down the higher part of our
natures, even our sense of right and wrong, then we
wonder how anybody of sane mind can become a victim
to its use.

TEA AND COFFEE.

These are so familiar to us all that it is only necessary to call attention to the fact that they are drugs, and are described in medical works as good for certain ailments.

The value of tea and coffee is doubted by many wise physicians. It is certain that most people would be better off without them.

Bad Effects sometimes. If the coffee be taken late in the day, it is likely to prevent sleep ; and the use of strong tea continued for a long time is likely to cause nervousness and dyspepsia.

Bad Effects on All. Neither strong tea nor coffee can be used for any length of time in large quantities without bad effects on the system. Their use by the young is likely to cause stomach troubles, headaches, nervousness, and wakefulness.

TOBACCO.

Its History. When Columbus, in 1492, landed on the Island of Hispaniola, he sent out his men in all directions to see what they could find in the new country. In a short time some of them returned to the ship and told him that they had seen men who appeared to follow a very singular practice. They carried tubes in their mouths, into which they would put some kind of weed, and setting fire to this, they would blow the smoke through their mouths and nostrils.

Origin of the Habit of Smoking. It is well for us to remember that the habit of using this weed comes to us from savages. The hideous signs of Indians and other fearful-looking objects seen in front of tobacco shops forcibly remind us of this fact.

Its Introduction abroad. In the year 1587 Sir Walter Raleigh sent a company of men to Virginia. When they returned they brought to him two vegetables, — the potato, and tobacco. Sir Walter procured a pipe, tried the tobacco, and soon became very fond of it ; but for some reason he kept its use a secret.

The story goes that one day he sent his servant for a pitcher of beer. The servant, who returned sooner than was expected, seeing smoke issue from the mouth and nose of his master, immediately dashed the beer into his face and rushed from the room shouting for help, saying his master was on fire inside, and fast burning up.

Efforts to stop its Use. Soon tobacco was being used to a large extent, and many of the wisest men of England foresaw some of the evils that would result from the use of such a powerful drug. James I., king of that country, therefore caused a heavy tax to be placed on tobacco, trying thus to limit its cultivation. He even wrote a book on its use, in which he said that smoking was a practice "loathsome to the eye, hateful to the nose, harmful to the brain, and dangerous to the lungs." A little later the Russian Government tried to prevent smoking, with the penalty of the loss of the nose if a person were caught using tobacco ; and the Sultan of Turkey punished with death all smokers or snuff takers. But the rulers of that day soon found that all they could say or do did not check the practice, which was steadily on the increase ; so they contented themselves with levying a tax, thus enriching themselves at the expense of those who persisted in the use of tobacco.

Nicotine. The acting principle in tobacco is called nicotine. This is a powerful substance, acting as a deadly poison to all forms of animal life.

In a work written by Dr. Wood for the use of medical men only, he says : " In its action on the system, nicotine is one of the most powerful poisons known. A drop of it in a concentrated form was found sufficient to kill a dog, and small birds perished at the approach of a tube containing it." The instance is given of a child eight years of age whose head was washed with the juice pressed from tobacco leaves. It was thought the application would cure some slight trouble of the scalp ; but, on the contrary, it caused the most severe pain, and in a few hours death followed.

Nearly all insects shun tobacco, hence it is used to protect furs and woollen garments from moths.

Effects on Vegetables. Tobacco seems to destroy all forms of vegetable life. If plants be placed in a room in which there is a strong odor of tobacco, as in tobacco factories, or in tobacco drying-rooms, they will gradually droop and die.

The General Verdict. All medical writers are agreed that nicotine is one of the most deadly poisons to all forms of animal and vegetable life, and also that it is the presence of this poison in tobacco that gives it its power over those who use it. •

Its first Effects. The first use of tobacco almost invariably causes sickness. The healthy body feels its effects most keenly, and extreme prostration, nausea, and vomiting follow. If its use be persisted in, however, these acute symptoms disappear, and the slower effects begin to appear.

Its Effects on the Young. The effects on the body of the adult vary greatly, according to the individual, his habits of life, etc. It is certainly true that tobacco has a much more severe effect on some than on others. It may be laid down as a rule, from which there is scarcely an exception, that its effects on the systems of all who use it before the system is well matured, say twenty-five years of age, are bad; while if its use is begun earlier in life, at ten or twelve years, the effects are much more serious.

The General Rule. As a rule, the younger the person using tobacco, the more active will its effects be. This will hold true until manhood is reached.

Especially bad for Boys. Other things being equal, a boy who smokes or chews tobacco before he is twelve or fourteen years of age must expect to be shorter in stature, weaker in muscle, and to show less fitness in his studies than his companion who does neither.

Wise Laws. The effect of tobacco on the boys in the public schools in Paris was shown by their pale faces, weak muscles, imperfect circulation of the blood, and consequent poor recitations. On account of this, the Minister of Public Instruction issued an order forbidding its use by the students.

Some of the earlier laws of New England were wise in this respect. One ordained that no person under twenty years of age should use tobacco without first obtaining a certificate from a physician to the effect that the tobacco would be good for him.

The use of tobacco is prohibited in the military and naval schools of the United States Government because of its bad effects on the young men who are to become sailors or soldiers.

It shortens Life. Dr. A. B. Palmer, who was a professor for over thirty years in the Medical College with which the writer is connected, said that "boys and young men who use tobacco lose one fifth of the enjoyment and value, and at least one tenth of the length, of their lives."

Exceptions do not constitute a Rule. It is nothing in favor of tobacco that some persons can use it without apparent bad effects, any more than it is an argument in favor of alcohol that some persons do not appear, at first, to be harmed by it.

We know two persons who are made very sick every time they taste honey ; yet it would be foolish for us to say, because of this, that honey makes everybody sick, and therefore should not be used. Such cases are the exceptions, not the rule ; they do not serve as guides for us at all.

General Effects. Tobacco lessens the natural appetite for food and injures digestion. It irritates the mouth and throat, and frequently causes the voice to become husky and coarse. It makes the body restless and nervous, causing a peculiar sinking sensation at the stomach, which strongly tempts the smoker to try the effects of some alcoholic drink. Moreover, it causes the sight to become weak, and gives rise to ringing, buzzing noises in the ears. It affects the action of the heart, making it beat unsteadily, and leading to dizziness and rushing of blood to the head. It disturbs the sleep with distressing dreams. It weakens the action of the muscles, causing them to tremble. All these and other ill effects may be easily prevented, simply by letting tobacco alone.

There are three classes of persons on whom its effects may be studied : —

First, those who do not find any, or only a few, of these effects immediately following its use. A very small number belong to this class.

Second, those who notice a large number of these effects. Only a few of them may be present at one time, but sooner or later a large number of them will be experienced. To this class belong a majority of those who use tobacco.

Third, those who find that even the use of a small quantity of this drug brings upon them all the more severe symptoms mentioned above. They are made ill not only the first time, but every time they attempt to use it in any form; it acts as a poison each time it is taken. This class is about as large as the first.

It is a Filthy Habit. Chewing or smoking is a filthy habit, especially the former; while the use of tobacco in any way imparts a strong, disagreeable odor to the breath and clothing, which would be condemned socially if so many were not addicted to it.

Statesmen condemn it. In twenty-nine States and in the District of Columbia there are stringent laws forbidding the furnishing of cigarettes and tobacco to boys under sixteen or eighteen years of age. What does such action show? It shows that these representative men of the nation saw there was a terrible danger at hand, and that the time had come for the remedy to be applied. Let us all declare the use of tobacco an illegal and injurious business, and never engage in it.

Cigarettes are especially harmful. Not a single fact can be brought forward in their favor.

.

CHAPTER XXXIII.

ALCOHOL AND LONG LIFE.

IT is a fact that those who never indulge in alcoholic drinks, and do not use any of the narcotic drugs, have a prospect of living longer than those who do use them. We have learned that some diseases are caused by alcohol, and physicians know that many diseases are made worse by it. If a person has led a temperate life he will be more likely to recover from a severe illness than one who has been intemperate.

Two of the largest life insurance companies in England insure persons who never use intoxicating drinks at much lower rates than they insure others. Officers of these companies tell us that a temperate person at the age of twenty may be expected to live forty-four years longer; while if one is intemperate at twenty, he may be expected to live only fifteen years more. Twelve presidents of life insurance companies testify that the use of intoxicating drinks tends greatly to shorten life. Dr. Willard Parker, one of the greatest physicians who ever lived, said in a public lecture that " one third of all the deaths in New York city were caused, directly and indirectly, by the use of alcoholic drinks."

A physician who has examined the causes of the diseases and deaths of thousands of persons has recently said : " Alcohol is a poison, and belongs on the shelf of the druggist with other poisons." He adds that users of strong drink are more subject to acute diseases than those who do not use them, and that these diseases are more likely to prove fatal to the former than to the latter.

We thus learn that the use of alcoholic drinks tends to shorten life, to make the body more subject to disease, and to render diseases more fatal. The testimony of Dr. Parker, which could be multiplied many times over by that of other eminent men, shows that if we hope to live to a good old age we must not indulge in any kind of strong drink.

The fact that a few drink and live to be old, is nothing in favor of the use of alcohol. Sometimes a man is either so very large or so very small that he is exhibited in a show; but all men must not be judged by these exceptions. It would be foolish to say that all men are either very large or very small just because it might be possible to find one or two of that description. So it may be possible to find a man who has taken strong drink for years, and yet lives longer than many of his temperate neighbors. But such cases are rare exceptions, and of no assistance to us. For one such case there are a thousand where persons have been killed by the use of liquor. There are but few exceptions to the rule, which is : Alcohol weakens all the powers of man, and greatly shortens his days.

CHAPTER XXXIV.

WHAT THE AUTHORITIES SAY.

WE might fill a large book with extracts from the writings of men who are eminent in all fields of work, showing the terrible effects of alcoholic drinks on the human body. We propose to quote a very few, that our young people may read what those who are most competent to judge have to say.

The International Medical Congress at Philadelphia in 1876 said : " Alcohol is not shown to have a definite food-value."

Dr. Willard Parker, one of the most eminent medical men this country has produced, says: " Alcohol is an irritant poison, having no place in a healthy system."

Some of the leading dispensatories define alcohol as a poison. Drs. Taylor and Orfila, authors of standard works on toxicology, class it as a poison, and it is declared to be such by leading physicians in this country and in Europe.

Mr. Locke, familiarly known as Petroleum V. Nasby, took great pains to visit a large number of leading physicians in order to obtain their opinions as to the effect of alcoholic drinks on the human body. He says :

14

"The report they present is simply terrible. The habit of drinking fastens itself on its victim, and daily becomes more and more the wretched man's master, clogging up his liver, rotting his kidneys, decaying his heart, and stupefying and starving his brain, fastening upon him rheumatism, erysipelas, and all manner of painful diseases, and finally dragging him down to the grave at a time when other men are in their prime of mental and bodily vigor."

The noted Sir Henry Thompson, of London, says: " I have no hesitation in attributing a very large proportion of the most painful and dangerous diseases which come under my care to the ordinary and daily use of the fermented drinks taken in a moderate quantity."

Dr. G. S. Howe and Dr. Willard Parker both testify that one half of the idiots are children of those who have been harmed by drink. Dr. Howe ascertained the parentage of three hundred idiots, and found that one hundred and forty-five of them were the children of habitual drunkards, and over three fourths the children of intemperate parents.

Professor McIntosh says that " five sixths of all who have fallen by cholera in England were persons of intemperate habits."

Dr. Adams, who is a professor in the Medical Department of the University of Glasgow, says: " Alcoholic drinks are one of the great predisposing causes of cholera. I would place the sign over every shop in the city where liquor is sold, **Cholera sold here**."

In France, during the last war with Prussia, it was

found that over one half of all the cases of insanity were from the use of alcohol.

In the lunatic asylum at Dublin "nearly one half of the cases were known to be caused by the use of alcohol alone."

In America the proportion is not so great; but even here it is surprisingly large. It is estimated that twenty of each one hundred insane persons are so afflicted as a direct result of the use of alcohol, and thirty-five of each one hundred as an indirect result, making in all fifty-five of each one hundred insane persons so afflicted on account of the use of some form of alcohol.

Dr. N. S. Davis, of Chicago, who is a teacher, writer, author, and physician of the very highest authority, a man well known at home and abroad, says: "After an experience of over fifty years, I still believe that there is no form of alcoholic drink either necessary or desirable for internal use either in health or in the various forms of disease; but health can be better preserved and disease better treated without any use of such drink."

Dr. A. C. Rembaugh, a prominent physician of Philadelphia, says: "I have no use for alcohol as a food, drink, or medicine, and I believe it is never used, in large or small quantities, without absolute harm to the one partaking of it."

Dr. William Pepper, also of Philadelphia, and President of the University of Pennsylvania, a teacher, author, and physician of the widest reputation, says: "The habitual use of alcoholic drinks by healthy persons is highly injurious, and involves the risk of developing serious disease."

Henry C. and Sara A. Spencer, principal and vice-principal of the Spencerian Business College, of Washington, say: " In our thirty years' experience in teaching more than fifty thousand young people, we have found the effects of tobacco to be premature age, shattered nerves, mental weakness, stunted growth, and general physical and moral degeneracy, and, therefore, *we now decline to receive into our institution any who use this noxious weed.*"

Dr. Willard Parker says: " Tobacco is ruinous in our schools and colleges, dwarfing body and mind."

In the District of Columbia two hundred and fifty-seven physicians, five hundred and twenty-four officers and teachers of the public schools, all the trustees of the public schools, and eighty-six pastors of churches, signed a memorial to the Fifty-first Congress, declaring the evil effects of tobacco, and urging the passage of a bill prohibiting the selling, giving, or furnishing of tobacco in any form to persons under sixteen years of age.

The eminent men composing both Houses of this Congress agreed with the petitioners, and the bill was passed. The President of the United States signed the bill. Eminent men condemn tobacco. Science condemns it. Experience condemns it. Let us all cast it out.

CHAPTER XXXV.

BEFORE THE DOCTOR COMES.

THE knowledge we have acquired of anatomy, physiology, and hygiene should be of great assistance to us in times of emergency, as when an accident renders it necessary that something should be done at once. To wait idly until the doctor comes may cost a life, while prompt and proper action will often greatly aid him, and may prevent a fatal termination to the accident; though some persons feel that they must do something, no matter what. Flurry often increases the danger, while a little cool judgment might greatly lessen it. To know what to do, and how and when to do it, is a great deal.

When calling a physician, always inform him of the nature of the accident. that he may bring with him all necessary appliances and remedies. Examine an injured person with great care, as rough handling may open a wound which has ceased bleeding, making it bleed afresh, or the rough handling may cause a broken bone to injure some of the soft tissues.

Unconsciousness. A person may be rendered unconscious from any one of a number of causes. Unconsciousness may be the result of an injury to the brain, or arise from alcohol or opium poisoning, or from loss of blood, or other causes. The ordinary method of detecting unconsciousness is by touching the eyeball : if there is complete insensibility, the eyeball will not move, nor will the eyelids. The means of restoring consciousness depend upon the cause of unconsciousness.

Fainting. A person who has fainted should be placed at once on his back, and the head kept as low as the body, and not raised until he has recovered. This should be done so that the blood may flow readily through the brain. To keep the head raised high, or to maintain the body in the erect posture, may cause a fatal termination to the attack. Dashing a small quantity of water on the face, and holding an open bottle of ammonia to the nose, will perhaps aid. All clothing about the neck should be loosened. Never give brandy or any form of alcohol in such cases; a cup of hot coffee is much better.

Intoxication. When insensibility arises from intoxication it is a more difficult matter to restore consciousness. As a rule, very vigorous measures are not successful, and are often harmful.

Loss of Blood. When a person has lost so much blood that he is unconscious, he should be placed on his back with his head low. After the bleeding has ceased, he should be kept as quiet as possible until the doctor comes. If the body becomes cold, hot flannels and bottles filled with hot water should be placed around it.

Shock. Sometimes persons become insensible in consequence of a blow, or through fright, or by a fall; and yet none of these causes may be severe enough to injure in any way the organs or tissues of the body. As a rule it is only necessary to place the person on his back and give him plenty of fresh air. If the head is hot, a cloth moistened with cold water may be applied to it.

Injury to the Head. If the unconsciousness comes from this cause, it is a very serious matter, and nothing can be done until the doctor decides just how serious the injury is.

Fresh Air. Always give injured persons plenty of fresh air. Do not crowd about unless you can be of use. Only those who can assist should stand near an injured man, or even remain in the same room.

Fire. The clothing may take fire, and produce results that may prove fatal in a short time. When we see a person on fire we should grasp the nearest rug, shawl, blanket, large cloak, or heavy curtain, and wrap it tightly about his body. After this is done, it is wise to roll him on the ground, which, with the blankets, will be likely to smother the flames. If no blankets are at hand, then the simple rolling on the ground may put out the fire.

Burns and Scalds. Great relief will be given in cases of burns or scalds by covering them with soft linen or cotton cloth that is saturated with a solution of common soda. Put a tablespoonful of common soda in a cup of water and stir thoroughly. Wet the cloth in this water, and place it upon the injured parts. This will generally relieve the sting and ache of the burn. If no soda is at hand, then try cream, or a thick coating of dry flour. A liniment of equal parts of sweet oil and lime water is very useful.

Sunstroke. In cases of sunstroke the skin and head are usually very hot, and there may be partial or complete unconsciousness. Remove the patient to a cool place, lay him on his back, with the head slightly raised, and apply cold cloths to the head, at the same time bathing the face and head with cold water.

Sprains. Sprains are oftentimes of a serious nature, and recovery from them is usually slow, sometimes leaving a stiff joint. This should be bathed in either hot or cold water, whichever gives the greater relief.

Fractures and Dislocations. If a bone is broken or thrown out of joint, or if this is thought to be the case, the injured part should be kept perfectly quiet, and the patient made as comfortable as possible. It is better to wait a few hours for the physician to come, than to attempt to set a broken bone, or to handle the injured parts to learn what is the matter.

Bleeding. Nothing is more alarming than the sight of flowing blood. To check it often requires prompt action; nay, promptness may save a life. Do not try to stop the bleeding by tying great quantities of

clothing around the injured parts; it will soak up
the blood, while the bleeding may continue beneath. If
the blood comes from an artery, it will flow in jets or
spurts; but if it arises from a vein, the stream will be
a steady one. If the bleeding comes from the surface
of the body, it may generally be checked by pressure.
This may be applied with the fingers; or if the wound
is on the surface over a bone, a piece of cloth may
be folded so as to make a small pad and held tightly
pressed to the wound by a bandage. If the bleeding
is slight, frequent applications of cold water bandages
will stop the flow. Bleeding from the nose is the most
common and the least dangerous of all hemorrhages.
It is generally sufficient to apply cold water to the forehead or back of the neck, and to remain quiet for a short
time.

Poisons. Whenever it is feared that poison has been
swallowed, it is safest not to wait until the doctor comes,
but to cause vomiting at once, and thus get the poison
out of the stomach before it is all absorbed. Vomiting
is easily caused by giving the patient a large cupful of
warm water in which there has been thoroughly mixed
a dessertspoonful of ground mustard. If this does not
cause vomiting in a few minutes, the dose may be repeated. A tablespoonful of alum dissolved in a pint of
warm water taken at once may be effective. Aid the
drugs by thrusting the finger down the throat. After
the vomiting it is best to give a glass of milk in which
are the whites of two eggs well beaten.

INDEX.

PRONUNCIATION OF WORDS.

A-ce′tous.
Al-bū′mi-nous.
Al′che-mists.
Al′ĭ-ment′a-ry.
A-năt′o-my.
A-or′ta.
Ar-te′ri-al.
Au′dĭ-tŏ-rў.
Au′ri-cle.

Bĭle.
Bron′chi-al.

Căp′il-la-ry.
Cär′ti-lage.
Cer′e-bellum.
Cer′e-brum.
Chȳle (kĭl).
Chȳme (kĭm).
Cor′pus-cle (kôr′pus-l).
Cu′ti-cle.

Dĕ-lĭri-um trē′mens.
Dī′a-phragm (frăm).

Ep′i-glŏt′tis.
Ex-pī′ra-to-ry.

Fē′mur.

Găs′tric.

Hȳ′gi-ēne.

In-spi′ra-to-ry.
In-tĕs′tĭne.

Jaun′dice.

Lăch′ry-mal (lak′rĭ-mal).
Lac′tĕ-al.
Lär′ynx (inx).
Lym-phăt′ic (lim-făt′ik).

Măr′rŏw.
Măs-ti-cā′tion.
Me-dŭl′la ob-lon-ga′ta.
Mī′cro-scope.
Mū′cous (kus).
Mŭs′cle (mŭs′sl).

Nĭc′o-tĭne.

Œ-sŏph′a-gŭs (ē sŏf′a-gŭs).

Pan′cre-as.
Per′i-ŏs′te-um.
Phăr-ynx.
Pū′pil.
Pȳ-lō′rus.

Rĕs-pi-rā′tion.

Sa-lĭ′va.
Skĕl′e-ton.
Skŭll.
Spī′nal.
Stĭm′ū-lant.

Thŏ-răc′ic.
Trā′che-a (trā′ke-a).
Tȳm′pa-nŭm.

Vén′tri-cle.
Ver′te-bra.
Vĭl′lus.

www.ingramcontent.com/pod-product-compliance
Lightning Source LLC
Chambersburg PA
CBHW030123030726
47498CB00007B/2515